LASERS AND ELECTRO-OPTICS RESEARCH AND TECHNOLOGY

OPTICAL SPECTROSCOPY

TECHNOLOGY, PROPERTIES AND PERFORMANCE

LASERS AND ELECTRO-OPTICS RESEARCH AND TECHNOLOGY

Additional books in this series can be found on Nova's website under the Series tab.

Additional e-books in this series can be found on Nova's website under the e-book tab.

LASERS AND ELECTRO-OPTICS RESEARCH AND TECHNOLOGY

OPTICAL SPECTROSCOPY

TECHNOLOGY, PROPERTIES AND PERFORMANCE

NICOLAE TOMOZEIU, PH.D.
EDITOR

nova publishers
New York

For permission to use material from this book please contact us:
Telephone 631-231-7269; Fax 631-231-8175
Web Site: http://www.novapublishers.com

Library of Congress Cataloging-in-Publication Data

ISBN: 978-1-63321-197-1

Library of Congress Control Number: 2014942199

Published by Nova Science Publishers, Inc. † New York

As editor I dedicate this work to my parents.

CONTENTS

PREFACE

A book of *Optical Spectroscopy: Technology, Properties and Performance* would take various forms of presentation and would address different topics, according to the aims of the authors and the audience to which is prepared. The science of Optical Spectroscopy has multiple and deep roots in physics, chemistry and materials' science, but also in electronics and computing science. Therefore Optical spectroscopy can be seen as an interdisciplinary area dedicated in the one hand to generate and built-up knowledge in the field of spectroscopy and in the other hand to be used as a tool for practical investigations. This book has been conceived for the use of specialists and persons who have a strong interest in material characterization and direct investigation of various processes.

The topics have been selected to supply the physical background needed to understand the main spectroscopic principles and, on the same time to reveal new potentiality for applications. Subjects like Reflectance Spectroscopy, Infrared Attenuated Total Reflection or Photoreflectance Spectroscopy overview the *background* of classical methods in spectroscopy in contexts of new applications or reveal possibilities in new domains as medicine, environmental investigations, etc.

When we discussed about *technology* in optical spectroscopy, in this book we presented know-how of various spectroscopic measurements' methods embedded in characterization of materials, devices or processes:

- Polariscopy is used to determine with high sensitivity the internal strains in mono-crystalline silicon;

- Electronic Spectroscopy of Diffuse Reflection is dedicated to reveal special properties of the atoms in the superficial layer of dispersed materials;
- Optical Emission Spectroscopy is committed to successfully characterize plasma deposition processes with direct and in real time monitoring the properties of the deposited thin solid films.

The chapters contained within this volume are contributed by recognized experts in the field who kindly collaborated with their know-how and expertise. Each chapter has a short introductory section that provides the reader with an overview of the subject. The sections that follow debate theoretical models and/or experimental spectroscopic analysis. Since the field of optical spectroscopy applications is truly large, direct connections between particular cases debated in chapter and new applications are made. One of our ambitions with this book has also been to contribute to better relating different techniques of optical spectroscopy (considering both the fundamental and the applied science) to high tech industrial applications and life conditions' improvements.

I am truly grateful to all authors for their outstanding collaboration. It was a pleasure working with this international team. It is my hope that this volume will become the preferred reference book for a large scientific community working in optical spectroscopy development and applications in industrial fields; in this way I am sure that the effort of my collaborators will be rewarded.

I would like to express my gratitude to all colleagues and Nova Publisher staff who contributed to the achievement of this book. I acknowledge the very efficient technical help received from them during the editing work of this book.

Dr. Nicolae Tomozeiu, Ph.D.
Senior Scientist R&D
Oce Technologies
The Netherlands
nicolae.tomozeiu@oce.com

In: Optical Spectroscopy
Editor: Nicolae Tomozeiu
ISBN: 978-1-63321-197-1

Chapter 1

OPTICAL EMISSION SPECTROSCOPY USED TO INVESTIGATE PLASMA DEPOSITION OF THIN SOLID FILMS

Nicolae Tomozeiu [*]
A&PS, R&D Océ Technologies BV, Venlo, The Netherlands

ABSTRACT

Depositions of thin solid films are essential for microelectronic and optoelectronic devices. Deposition systems based either on plasma decomposition of a carrier gas or ion sputtering of a target, are used daily to deposit thin films of semiconductors / dielectrics / insulators. The properties of the deposited layers are determined by the plasma conditions. Using various tools to diagnose the plasma in real time during the deposition is crucial in assuring the reliability of the deposition and in obtaining the material with desired physical properties. Optical emission spectroscopy (OES) method is such a tool. In this work we consider several plasma types used in sputtering or chemical vapor deposition and reveal the main mechanisms in plasma by studying the OES spectra. Correlations between intensity of different lines in OES spectra and process parameters of the deposition are presented. The method of OES is here-presented for the SiO_x deposition by rf magnetron sputtering and for the a-C:H thin films grown by plasma enhanced chemical vapor deposition. Moreover, physical properties of these materials (SiO_x, a-

[*] E-mail address: Nicolae.Tomozeiu@Oce.com.

C:H) are correlated to the OES signals intensities. In this way, the OES method is described as a toll to investigate plasma during the deposition and to tune directly and in real time the process of deposition.

Keywords: Optical Emission Spectroscopy, CVD, sputtering deposition, thin dielectric films

1. Introduction

Thin solid films of various materials (metal, semiconductor or dielectric) are used in industrial applications having a well defined technical role in the device design or functionalizing the surface properties. In many cases these layers are so thin that their presence is unnoticed unless we undertake an analysis of the device functionality. Today the technology of solar cells, flat panel screens, last generation of television apparatus, i-pads and all kinds of phones with performances unthinkable few years ago, all nano-technologies are pushing the limits of three dimensions to almost two dimensions like systems. And here the thin films and the technologies to produce them play an important role. For specialists the most used methods to obtain these thin films are the so-called plasma deposition techniques, where energetic particles from plasma are used to break the chemical bonds of atoms / molecules with their neighbors and making them free with a certain kinetic energy that allows them to reach the deposition substrate.

There is a big variety of plasma sources that has been developed for depositions of such layers; as examples we mention the rf- [1], dc- sputtering [2], or chemical vapor deposition [3], [4], [5]. As a common characteristic, the plasma in almost all these deposition techniques is mainly based on inert gasses (He, N_2, Ne, Ar, Kr).

In this work we consider two classes of materials deposited as thin films using plasma techniques: silicon oxide deposited by rf sputtering and hydrogenated amorphous carbon obtained by plasma enhanced chemical vapor deposition (PECVD).

Thin silicon oxide films are present in almost all microelectronic and optoelectronic integrated circuits. In specific applications that require thin films with particular electrical and optical properties silicon suboxide (SiO_x with $0<x<2$) films are used instead of silicon dioxide (SiO_2) [6]. The deposition technology for such layers is part of the device technology and must be compatible with it. In many applications low temperature depositions are necessary and sputtering technique is very appropriate. Energetic charged particles from plasma (e.g., ions) bombard a polycrystalline silicon target and sputter silicon atoms and/or ions. The composition of the deposited layer depends on the sputtering rate and on the amount of oxygen present in the deposition reactor. The deposition conditions (rf power, pressure, gas flow, etc.) have a great influence on the structure, composition and physical properties of the deposited material. Studies on the structural, chemical-physical and mechanical properties of these layers have been made and many theoretical and phenomenological aspects are well understood [7-10]. Thus, the electrical and optical properties are strongly related to the material structure. The Infrared Spectrometry (IR), X-ray Photoelectron Spectroscopy (XPS) and Electron Paramagnetic Resonance (EPR) have shown the structural and compositional effects on the defect density distribution and valence band shape [11]. But all these parameters have been measured post-deposition.

Semi-transparent amorphous hydrogenated carbon (a-C:H) films have been studied in the last decennia because of their potential technological advantages. Physical and chemical properties such as mechanical hardness, chemical inertness, wear resistance, bio-compatibility and infrared transparency are the main promoters of interest for these layers [12], [13]. As applications where a-C:H thin films are used we mention: micro-electromechanical systems [14], [15] optically activated field effect transistors [16] and detectors for dosimetry applications [17]. The PECVD deposition method is one of the most used to obtaining a-C:H films because of its versatility [18]. Here the plasma is used to decompose the molecules of a gas that contains carbon and hydrogen atoms; with their kinetic energy, these atoms are deposited on the substrate as well as on the PECVD reactor walls. The properties of the deposited a-C:H depend on this energy and on the processes at the surface of the just deposited material. We mention that a large number of deposition parameters (rf power, gas pressure and composition, magnetic field, substrate temperature, etc) are involved in obtaining thin layers with the desired properties.

An important step is made when the properties of the deposited film can be tuned during the deposition by varying the deposition conditions. In other words, the relation between the gas phase, which is the plasma region in the deposition vessel, and the physical properties of the deposited films is very important. Using optical emission spectroscopy (OES) measurements that study the emitted spectra of the atomic species from plasma, such correlations can be established in their complexity considering the deposition conditions.

In this work we present the OES technique used to study the plasma during the deposition of SiO_x layers by rf sputtering and a-C:H films by PECVD revealing correlations between the intensity of the OES signal and physical properties of the deposited layers. Section 2 is dedicated to describing main processes in plasma used for depositions of semiconductor / dielectric thin films. Then the method of OES is described in section 3 titled "spectroscopy with plasma light". Section 4 is dedicated to a general presentation of an OES experimental set-up and particularization for rf magnetron sputtering depositions of silicon oxide and PECVD depositions of a-C:H. From gas phase (plasma) to thin solid layers is the subject of the last paragraph in this section.

2. PROCESSES IN PLASMA

Plasma – a phase of the gas matter where a significant percentage of the atoms/molecules is ionized - is generally created via electrical discharge of the gas between two electrodes. The electrical field that generates the discharge can be ac (alternating current) frequency or dc (direct current). The most used ac discharge is in the rf (radio-frequency) domain: as 13.56 MHz. The gas used to produce plasma is, in general, chemically inert (e.g., argon). In special applications: reactive sputtering or chemical vapor depositions, etc., reactive gasses as oxygen, nitrogen, or complex molecules as methane, acetone, silane, etc that contains the chemical element needed to be deposit, are introduced in the deposition chamber, in the plasma volume. Ionization of these reactive atoms / molecules via discharge makes these particles to interact with other atoms present in the deposition reactor. In plasma there are neutrals, ions and electrons. The latter are very small in dimension and mass when compared to the ions; this means that they have a great mobility and follow the oscillations of the rf electrical field. The ions, due to their mass, are not able to follow the field. As a consequence, the plasma volume is more positive than any object it is in contact with. In this way, a thin sheath region between plasma and substrate is formed. Because the thickness of this sheath region is very small, the electrical field can reach large values. Any ionized particle (atom or molecule) that diffuses to the edge of the sheath region interacts with this electrostatic field and is accelerated towards the neighboring surface. A classical design for plasma generation is shown in Figure 1, where a system with parallel electrodes is considered. The gas pressure inside the deposition chamber is established by the pressure of the inlet gasses (e.g., $Ar + O_2$) and the pomp speed of the gas evacuation. In the plasma there are neutrals, ions, electrons and molecular entities. All surfaces exposed are subject to an energetic ion bombardment from plasma. This has an effect on the temperature of those surfaces and, of course, on the deposition process that occurs on the substrate. When the plasma is characterized by a low fractional ionization, the electrons can be maintained at very high equivalent temperatures, while the neutral atoms remain at the ambient temperature. These energetic electrons can induce dissociation of precursor molecules and creation of large quantities of free radicals. As a function of the plasma condition, the deposited films can be exposed to energetic ion bombardment which leads to increasing the film density, changing its mechanical and electrical properties (see in the next sections for both the SiO_x and a-C:H layers) of the film.

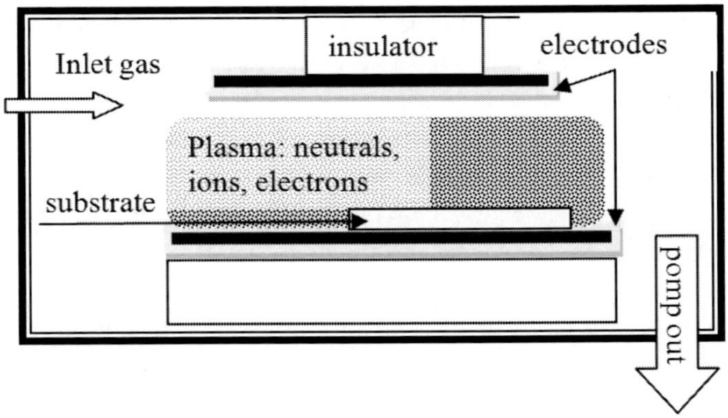

Figure 1. A schematic representation of a deposition setup with plasma formed between two parallel electrodes. The inlet gas can be Ar +reactive gas and the power supply can be rf or dc.

In order to briefly review some of the physical processes in the plasma volume, let's consider a plasma of argon. These atoms can be subjected to:

a) Ionization and excitation by electron collisions

Processes as electron excitation or ionization impacts in Ar self-sustained plasma are described by reactions like:

$$Ar + e \rightarrow Ar^{(*,+)} + n \cdot e \tag{1}$$

where Ar^* and n=1 describe the excitation phenomena and Ar^+ and n=2 depict the ionization. In the last process, a multiplication of the available electrons occurs.

What is the difference between the ionization or the excitation process? If the argon atom is in the ground state, we can speak of *direct electron impact ionization*, when the energy of the electron implicated into collision with Ar atom is minimum equal to the ionization potential of the argon atom, 15.76 eV. But the ionization could also be the result of *two-steps electron impact* process, if the argon atom is in a status characterized by the so-called metastable levels (11.55 or 11.72 eV). For this process the electron has to have the minimum energy of 4.04 eV to be able to overpass the ionization potential.

The physical parameter that characterizes the interaction between two particles (electron and argon atom) is the cross-section, which is dependent on the electron energy; for these ionization processes the cross-section parameter reaches the maximum of $3{\cdot}10^{-16}$ cm^2 and $8{\cdot}10^{-16}$ cm^2 around 80 eV and 10 eV, respectively [19].

If the energy transfer between the electron and Ar atom is so small that the ionization does not occur, there is a great chance that the so-called *electron impact excitation* process occurs. This means that one electron of the Ar atom is excited to a higher energy level with the atom. By de-excitation, the light so much needed for optical emission spectroscopy measurements is produced. This process is energy dependent, and the minimum energy required is 11.55 eV; the cross-section of the excitation process has the maximum value of about $1.6{\cdot}10^{-16}$ cm^2 for energy of about 20 eV [20].

b) Ionization and excitation by collisions with energetic argon ions

Processes of ionization and/or excitation are the result of collisions between argon atoms with fast argon ions or with argon atoms in metastable states. The mathematical representations for these processes are:

$$Ar + Ar_{fast}^{+} \rightarrow Ar^{(*,+)} + Ar_{fast}^{+} + n \cdot e \qquad (2)$$

where n=0 for the excitation process (the collision result is Ar^*) and n=1 for the ionization process (it results in Ar^+).

The cross-section values of these processes are also energy dependent. The maximum values, comparable with the value of the electron impact ionization and excitation are reached for energies of the fast ions higher than 1000 eV. Such energy values could be gained by the argon ions in the plasma sheath near the cathode. Highly energetic argon atom could be the result of a reflected argon ion on the cathode surface after its neutralization. Therefore, the ionization/excitation processes described by relation (2) are important only when taking place next to the cathode [21].

Before getting into another subject, we mention that there are also other mechanisms for argon ionization, respectively excitation; the most considered mechanisms are:

a) *thermal ionization/excitation,* which is the result of a collision between an argon atom and atoms at "thermal energy" (the latter can be either argon or atoms from the plasma vessel walls);
b) *photo-ionization* is the process where an Ar atom interacts with an energetic photon and is ionized. The cross-section for such an event is $3.7 \cdot 10^{-17}$ cm^2 at 15.8 eV and this value rapidly decreases with increasing the photon energy. The *photo-excitation* is a similar process, only the result is the excitation of the argon atom on a superior level of its electronic energy.

However, it has to be noted that these processes have a minor significance compared with electrons and fast particles impact ionization and excitation.

What happens with molecules/atoms of an adjacent gas introduced in plasma either for decomposition or as a reactive gas?
The easiest answer to this question is that electrons, ions, neutrals and photons from plasma will collide with such atoms / molecules and will exchange energy. Considering the energy of these plasma constituents and taking into accounts the conservation laws for energy and momentum, excitation and/or ionization are the expected physical processes. Of course the ionization and excitation energies are different for atoms/molecules like Si, O_2, N_2, CH_4, C_2H_2, SiH_4, etc.

In order to exemplify, let's consider plasma of pure oxygen (as we have considered the argon plasma). The collision between an oxygen molecule and an energetic electron from plasma can have the following consequences [22]:

a) dissociation

$$O_2 + e \rightarrow O + O + e \qquad \text{(a)}$$

or

$$O_2 + e \rightarrow O + O^- \qquad \text{(b)} \qquad (3)$$

where O^- is a negative oxygen ion (the oxygen atom accepted one electron on its last electronic energy level);

b) excitation

$$\begin{gathered} O_2 + e \rightarrow O^* + O + e \qquad \text{(i)} \\ \text{or} \\ O_2 + e \rightarrow O_2(a^1\Delta_g) + e \\ \text{and more} \\ O_2(a^1\Delta_g) + e \rightarrow O^* + O + e \qquad \text{(ii)} \end{gathered} \tag{4}$$

where $O_2(a^1\Delta_g)$ known as singlet delta oxygen [23] is characterized by a low excitation energy of 0.98 eV and O^* is the excited state of an oxygen atom. The singlet delta oxygen is important to produce the atomic oxygen via equation (ii) of relation (4), in plasmas used for depositions of oxides of various elements (e.g., silicon). Although this is an important channel to obtain atomic oxygen, in [22] it was shown that the reaction probability is small in comparison with other channels in plasmas used in such depositions.

Electron impact excitation of ground-state atomic oxygen has as result an excited atom:

$$O + e \rightarrow O^* + e \tag{5}$$

that leads by de-excitation to light emission at 844.6 nm.

Therefore, impacts of various particles from plasma with a certain type of atom specie will excite the atom on a meta-stable energy level that via de-excitation generates light with a specific wavelength that can be collected from plasma volume and studied. The rates of excitation – de-excitation processes are known and the number of detected photons can be correlated to the atom concentration of that specie in plasma.

3. Spectroscopy with Plasma Light

As it was shown in the previous section, atoms and molecules from the plasma volume are subject of collisions with similar particles and/or electrons. These interactions can be inelastic or elastic. For us the inelastic collisions are of interest because they result in either excitation (an electron from the atomic shell is placed on a superior energy level) or ionization (the electron is completely out from the atomic structure). As it was already mentioned, the excited states are metastable, and phenomena of de-excitation are generally accompanied by *photon emission*. This gives the color of the plasma bulk. The wavelength of the emitted photons is a signature of the atom species and is used for identification of the various species present in plasma.

Such an optical emission is characterized by intensity, shape of line/ band (for atom/ molecule), line broadening and position in frequency range (scale). The optical spectroscopy uses the light collected from plasma and studies these parameters. On the other hand, plasma parameters as: electron density and electron temperature, species density and their temperature values have a great influence on the emitted spectrum. Therefore a correlation between the plasma parameters and the properties of the plasma light is of interest. It is easy to foresee that the intensity of a certain emitted radiation depends on the density of the particles involved in the excitation mechanism and on their energy. About the last parameter we mention that in many types of plasma the electrons are highly accelerated within the electromagnetic field from plasma and gain a high energy. This is characterized by the electron temperature, T_e. When electrons collide with atoms and/or molecules, the gain in kinetic energy for these large entities is so small, that practically the parameter characterizing the energy of the entire system – the gas temperature, T_g – changes much less. However, the species temperature T_g influences the shape of emitted peak/ band from the spectrum, while the density of the atomic/molecular specie determines the line intensity and its broadening.

In a simple mathematical model, the intensity of the emitted light by an atomic specie as a result of impact with electrons is:

$$\Im(A, \lambda_{ij}) = \Psi(A) \cdot k(T_e) \cdot n_e \cdot n_A \qquad (6)$$

where $\Psi(A)$ is a function that contains parameters specific to the specie A and characteristics of the measurement system, $k(T_e)$ is the rate coefficient, n_e is the electron concentration in plasma and n_A is the concentration of the atomic specie A. The rate coefficient $k(T_e)$ considers the electron energy distribution f(E) and the cross-section $\sigma(E)$ for the collision event:

$$k(T_e) = \sqrt{\frac{q}{m_e}} \int_{E_{th}}^{\infty} f(E) \cdot \sigma(E) \cdot \sqrt{E} \cdot dE \qquad (7)$$

and defines the "efficiency" of the collision. The electron is characterized by its mass, m_e, and electric charge, q. These rate coefficients are very important for understanding the OES spectra. The effective cross-section can be found in literature as a function of electron temperature. For Ar plasmas, the cross-section $\sigma(E)$ of the electron-Ar atom collisions has been analytically reported by Phelps [24]. The energy distribution of electrons in plasma is generally described by Boltzmann distribution. Both the electron temperature T_e and electron density n_e can be measured via other experimental methods (e.g., Langmuir probe).

Considering relation (6) there are all necessary elements to estimate the relative intensity of the certain emitted wavelength; it is relative because the function $\Psi(A)$ has to be evaluated considering the experimental conditions and the atomic specie, A. In general this is solved either via a calibration measurement or the results are presented relative to a peak intensity considered as reference.

It is necessary to mention that, with a certain probability, the ions from plasma can get excited and emit light, too. The specific wavelength is slightly different from that of the atom counterpart. This means that with a good resolution spectrometer, the ionization degree in the plasma could be also determined via OES. In order to recognize the "light source", in OES a special notation is used: Ar I means a line from a neutral argon atom, Ar II means a line from a single-ionized argon, Ar III is a line from a double-ionized argon atom, etc.

4. Experimental Set-up

An OES set-up consists of an optical fiber that collects light from plasma and a spectrometer that analyzes this light and reveals the optical spectrum of the emitted light. The optical fiber can be mounted inside the main deposition chamber and via a special feedthrough at the wall of the deposition reactor it reaches the spectrometer controlled by a computer. Also, it is possible that the plasma light is collected by the optical fiber via a quartz window specially mounted in the reactor wall, therefore from outside. All these are experimental details that will influence the function Ψ from relation (6). A schematic connection between the spectrometer of the OES system and a sputtering deposition system is shown in Figure 2. The optical fiber is mounted inside the deposition system with a device to avoid material deposition on its active section. By moving the optical fiber inside the deposition chamber, different sections from the plasma volume can be monitored. It is important that the spectrometer is chosen such that it covers the whole spectral range of the emitted light, especially for the atoms of interest.

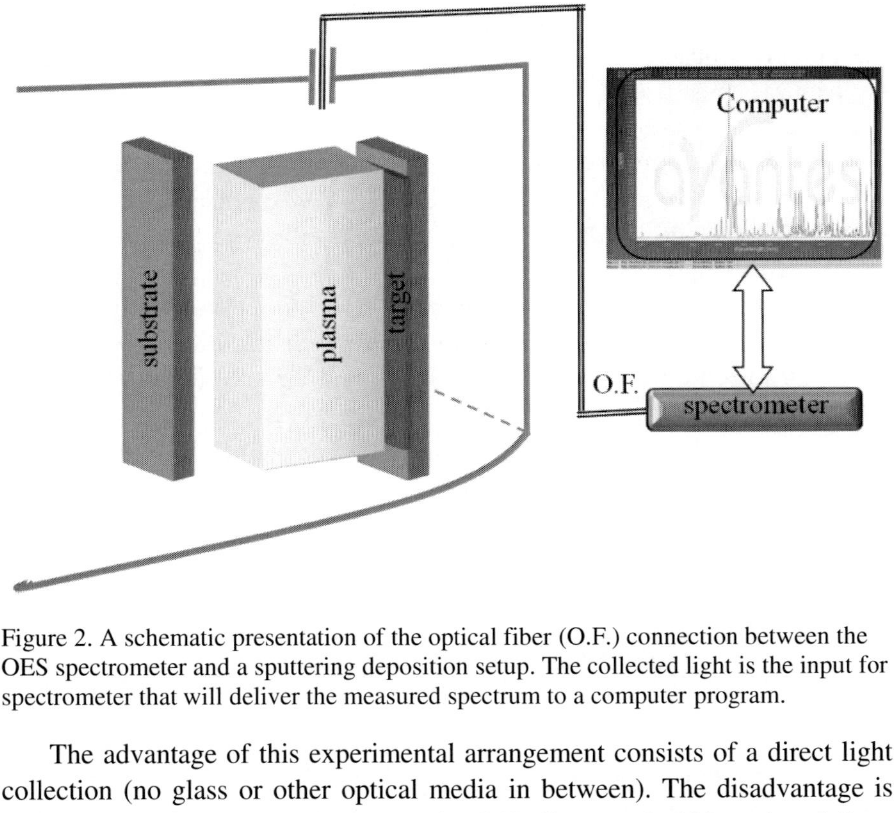

Figure 2. A schematic presentation of the optical fiber (O.F.) connection between the OES spectrometer and a sputtering deposition setup. The collected light is the input for spectrometer that will deliver the measured spectrum to a computer program.

The advantage of this experimental arrangement consists of a direct light collection (no glass or other optical media in between). The disadvantage is related to the small collecting area (the O.F. diameter is 200 μm) and from here, a small output signal. The spectrometer has options to select measurement conditions in a way that such situations are solved. With proper calibration, the results are very good. The integration time used to obtain an experimental point and the possibility to average the measurements over a variable number of spectra enlarge the flexibility and the accuracy of the OES method.

4.1. OES on Plasmas Used to Deposit SiO_x Thin Films by rf Magnetron Reactive Sputtering

The reactive sputtering via a rf magnetron was successfully used to deposit thin films of silicon suboxide SiO_x (with $0<x<2$) and other dielectric materials. The system itself was described elsewhere [25], [26], and more about the SiO_x material can be found in [11] and [27]. Briefly, hereunder are few details:

- SiO_x layers were deposited in a sputtering system equipped with a 13.56 MHz r.f. magnetron in a controlled oxidizing environment; silicon atoms have been sputtered from a polycrystalline silicon target using Ar plasma;
- the deposition chamber was pumped down to a base pressure of less than 5 10^{-5} Pa before the plasma gas (argon) was introduced. During deposition, the pressure was 0.56 Pa and the substrate was at room temperature;
- in order to get SiO_x layers with different oxygen content, the oxygen flow introduced in the vessel was varied between 0 and 20 sccm;
- the distance between target and sample holder was 6 cm;
- the deposition rate varies with the oxygen content, from 5.6 nm/min for a-Si (x=0) to 12 nm/min for SiO_2 (x=2);
- the composition of the deposited layers was determined by Rutherford backscattering spectrometry (RBS) and infrared spectrometry (IR).

The OES system was based on an Avantes AVS-SD2048 multi-channels spectrometer with an entrance for the optical fiber. Gratings of 1200 lines/mm provided a resolution of 0.17 nm on the investigated wavelength region: 200–874 nm. The light detection was assured by a CCD linear array. Each spectrum resulted by averaging 10 spectra, each of them being collected with an integration time of 50 ms.

As a result of collisions in plasma, an atom/ion (argon, silicon or oxygen) can receive enough energy from its partner in collision so that one of its electrons is excited from a lower energy state, E_{lower}, to a higher one, E_{upper} (the upper state, in Table 1). Generally this state is metastable and the electron on this level has a very low lifetime. By de-excitation the atom/ion emits light characterized by a wavelength λ:

$$E_{upper} - E_{lower} = h \cdot \frac{c}{\lambda} \quad (8)$$

where h is the Planck's constant and c is the light speed.

Analyzing light from such plasmas, the spectrum is formed from lines assigned to argon (more than 65), oxygen (two lines) and silicon (three lines) atoms. Part of them is given in Table 1 as electronic transitions and their generated wavelengths according to the NIST Atomic Spectra Database [28].

There is a probability that the emitted photon collides with an atom in the E_{lower} energy state and the exchanged energy is enough to excite it again. This phenomenon is called *self-absorption* and it is measured by effective absorption cross-section:

$$S_L = 0.33 \cdot \lambda^3 \sqrt{\frac{M}{T} \cdot \frac{g_j}{g_i} \cdot A_{ij}} \quad (9)$$

Table 1. The principal lines from plasma used in reactive magnetron sputtering of SiOx and the atomic states involved

OES line	Wavelength (nm)	Upper state (j)	Lower state (i)
Ar I	420.07	$3s^23p^5(^2P_{3/2})5p^2[5/2]_3$	$3s^23p^5(^2P_{3/2})4s^2[3/2]_2$
Ar I	696.54	$3s^23p^5(^2P_{1/2})4p^2[1/2]_1$	$3s^23p^5(^2P_{3/2})4s^2[3/2]_2$
Ar I	706.72	$3s^23p^5(^2P_{1/2})4p^2[3/2]_2$	$3s^23p^5(^2P_{3/2})4s^2[3/2]_2$
Ar I	738.4	$3s^23p^5(^2P_{1/2})4p^2[3/2]_2$	$3s^23p^5(^2P_{3/2})4s^2[3/2]_1$
Ar I	750.39	$3s^23p^5(^2P_{1/2})4p^2[1/2]_0$	$3s^23p^5(^2P_{1/2})4s^2[1/2]_1$
Ar I	763.51	$3s^23p^5(^2P_{3/2})4p^2[3/2]_2$	$3s^23p^5(^2P_{3/2})4p^2[3/2]_2$
Ar I	772.38	$3s^23p^5(^2P_{3/2})4p^2[3/2]_1$	$3s^23p^5(^2P_{3/2})4s^2[3/2]_2$
Ar I	794.82	$3s^23p^5(^2P_{1/2})4p^2[3/2]_1$	$3s^23p^5(^2P_{1/2})4s^2[1/2]_0$
Ar I	811.53	$3s^23p^5(^2P_{3/2})4p^2[5/2]_3$	$3s^23p^5(^2P_{3/2})4s^2[3/2]_2$
Ar I	842.46	$3s^23p^5(^2P_{3/2})4p^2[5/2]_2$	$3s^23p^5(^2P_{3/2})4s^2[3/2]_1$
O I	777.42	$2s^22p^3(^4S^o)3p\ ^5P_2$	$2s^22p^3(^4S^o)3s\ ^5P_2$
O I	844.63	$2s^22p^3(^4S^o)3p\ ^3P_2$	$2s^22p^3(^4S^o)3s\ ^3S^o{}_2$
Si I	251.61	$3s^23p4s\ ^3P_2$	$3s^23p^2\ ^3P_2$
Si I	288.16	$3s^23p4s\ ^1P_2$	$3s^23p^2\ ^1D_2$
Si I	390.55	$3s^23p4s\ ^1P_1$	$3s^23p^2\ ^1S_0$

where M is the atomic mass of the element, T is the temperature, g_i is the Lande's factor for levels involved in the optical transition and A_{ij} is the Einstein coefficient for optical transition between the energy levels i and j. Calculations of the effective absorption cross-section for Ar, Si and O atoms are shown in Figure 3.

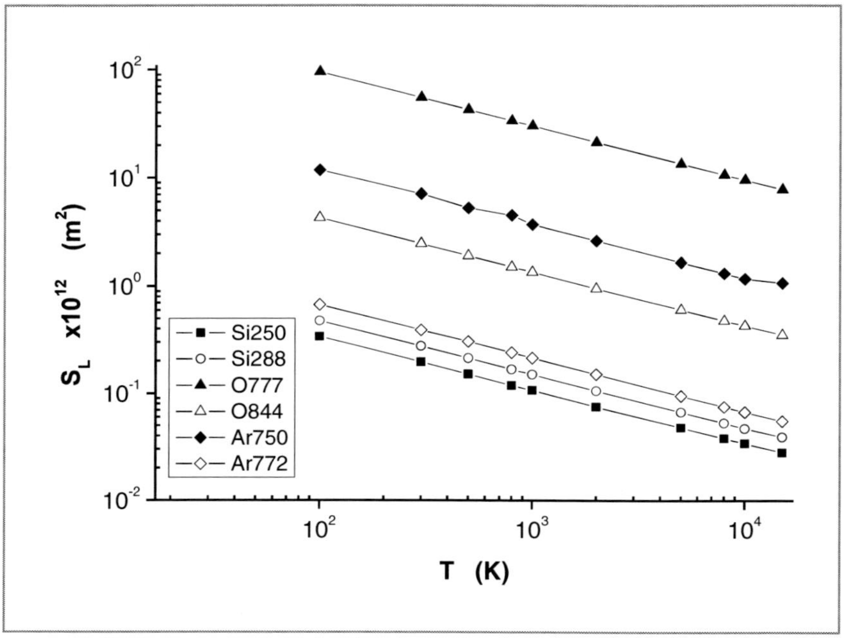

Figure 3. The effective absorption cross-section values that characterize the self absorption for some wavelengths.

According to these data, the self-absorption for silicon (Si250) is smaller than that for oxygen (line O777nm) and argon (line Ar750nm). In many scientific papers the self-absorption is neglected. Between Si and O effective absorption cross-section values there are 2 orders of magnitude.

In Section 2 possible interactions between electrons, neutrals and ions in plasma of argon-oxygen mixture used to sputter silicon have been presented. In Table 1, a number of 15 lines identified in the OES spectrum are shown. The strongest line for argon, λ=750.4 nm, is a result of the electron excitation impact with argon atoms. Under this assumption, considering the definition given in (6), the line intensity can be modeled as:

$$\Im_{Ar750} = \alpha_{Ar} \cdot n_{Ar} \cdot n_e \cdot \int_E f_e(E) \cdot \sigma(E) \cdot dE = \alpha_{Ar} \cdot n_{Ar} \cdot n_e \cdot k_{Ar}(\lambda) \quad (10)$$

where n_{Ar} is the particle density of the unexcited argon, n_e is the free electron density in plasma, $f_e(E)$ is the electron energy distribution function, $\sigma(E)$ is the excitation cross-section and α_{Ar} is a constant that depends on the spectral response of the spectrometer. The excitation rate coefficient, k_{Ar}, is defined by the integral function from relation (10).

4.1.1. OES for Sputtered Silicon Atoms

One question that immediate arises is: *what about the excitation of the silicon atoms?*

The important excitation mechanisms for silicon atoms are the collisions with electrons and argon atoms. Considering the line intensity described by relations similar to (10), the silicon intensity OES line can be written as:

$$\Im_{Si} = k_{Si\_e} \cdot n_{Si} \cdot n_e + k_{Si\_Ar} \cdot n_{Si} \cdot n_{Ar} \quad (11)$$

where n_{Si} is the particle density of the unexcited silicon atoms.

It is important to understand which excitation mechanism of the two is preponderant. On the first approximation, the pressure value inside the deposition chamber is proportional to the argon density number and the silicon density number is proportional to the deposited silicon amount: that is, $n_{Ar} \sim p$ and $n_{Si} \sim [Si]_{dep}$. The amount of deposited silicon atoms can be determined via RBS or optical measurements. Under these assumptions, the ratio between the OES silicon intensity and the deposited silicon amount is:

$$\frac{\Im_{Si}}{[Si]_{dep}} = \frac{k'_{Si\_e}}{k'_{Ar\_e}} \cdot \frac{\Im_{Ar750}}{p} + k'_{Si\_Ar} \cdot p \quad (12)$$

This relation is very important because a plot of $\Im_{Si}/[Si]_{dep}$ versus the pressure will point out the most important excitation mechanism. Also, the graphic $\Im_{Si}/[Si]_{dep} = f(\Im_{Ar})$ will offer the alternative checking solution. In Figure 4 are shown the results of this model applied to one of the two lines emitted by silicon atoms for plasmas without oxygen. The proportionality with pressure of the experimental data shows *that the most efficient excitation mechanism of Si atom is the collision with the neutral argon.*

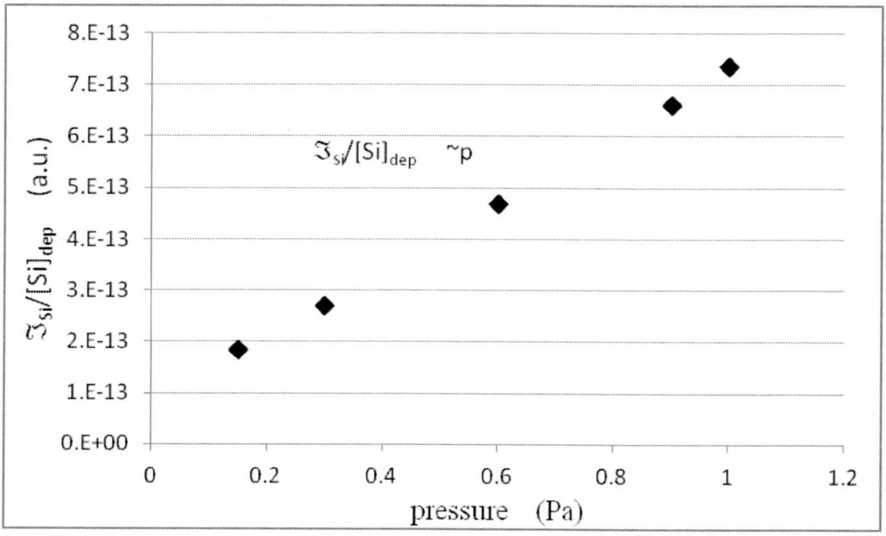

Figure 4. The variation of the $\Im_{Si}/[Si]_{dep}$ ratio with the pressure gives indication about the silicon excitation mechanism, according to relation (12).

4.1.2. The Provenience of the OES Oxygen Line

In order to obtain silicon oxide via this reactive rf magnetron sputtering it is necessary that the sputtering of the silicon atoms is made into an ambient with molecular oxygen. The measured OES spectrum in oxygen plasma (only oxygen!) has revealed bands with small intensity assigned to the molecular oxygen [29]. Adding argon in the gas plasma this signal cannot be distinguished from the noise level. But another oxygen signal appears at 777.4 nm assigned to atomic oxygen. The excited state of the atomic oxygen can be obtained by electron impact with the atomic oxygen in the ground state, or by so called *the dissociative excitation*, when the electron impact with molecular oxygen produces an excited atomic oxygen (see relation (4)). Considering the two sources for the oxygen emission, the model described by (10) for the line intensity will produce:

$$\Im_{O\_777} = \alpha_O \cdot k_{O\_e} \cdot n_O \cdot n_e + \alpha_{O_2} k_{O_2-e} \cdot n_{O_2} \cdot n_e \qquad (13)$$

where the amount of atomic oxygen is unknown in the vessel. The electron density is also unknown without a Langmuir probe measurement. But, the argon intensity line is known (see relation (10)). The ratio between the intensity of oxygen and argon lines will lead to:

$$\frac{\Im_{O\_777}}{\Im_{Ar\_750}} = \alpha' \cdot \frac{k_{O\_e}}{k_{Ar\_750}} \cdot \frac{n_O}{n_{Ar}} + \alpha'' \cdot \frac{k_{O_2\_e}}{k_{Ar\_750}} \cdot \frac{n_{O_2}}{n_{Ar}} \quad (14)$$

This relation could provide criteria to identify the preponderant mechanism to obtain excited atomic oxygen and, from here, the emitted light at λ=777 nm. A representation of $\Im_{O\_777}/\Im_{Ar\_750} = f(n_{O_2})$ is given in Figure 5 for plasmas produced with 94 sccm of Ar plus the added oxygen according to the O_2 flow axis. The linear dependency could reveal that the mechanism of molecular dissociation is implicated. A more elaborated discussion on this subject is given in [30].

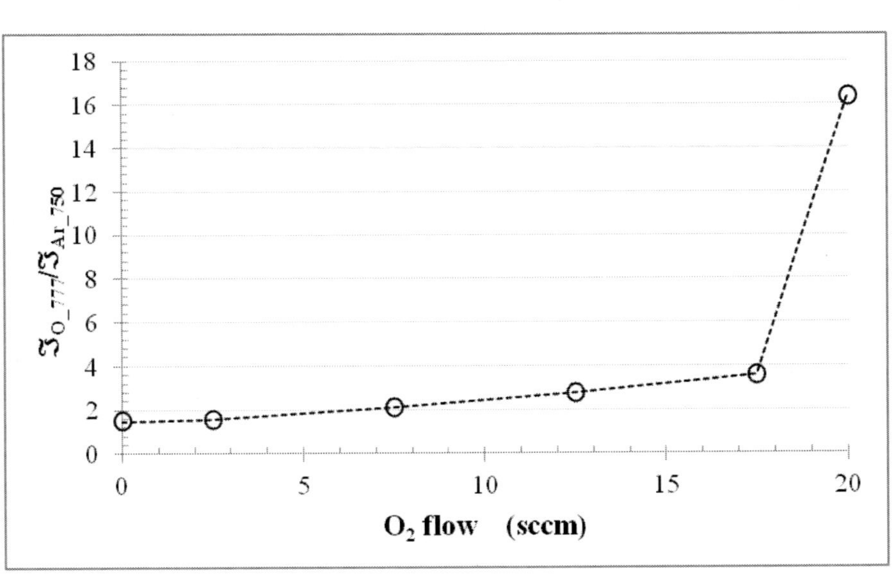

Figure 5. A linear dependency in the $\Im_{O\_777}/\Im_{Ar\_750} = f(O_2)$ plot shows that the dissociative excitation processes of the molecular oxygen are the most probable sources for oxygen OES, λ=777 nm.

Important to mention is the last experimental point from Figure 5 which is completely out of the trend obtained with smaller oxygen flows; for larger amount of oxygen (reactive gas) in the deposition vessel, plasma gets in the so-called "oxidic mode" and the deposited material is almost SiO_2.

4.1.3. From OES Data to the Composition of the Deposited SiO_x Layer

With the OES signals assigned to oxygen (λ=777 nm) and silicon (λ=250 nm) it is possible to define a new parameter called "x_{plasma}" similarly to "x" parameter from SiO_x deposited sample: $x_{plasma} = \Im_{O\_777} / \Im_{Si\_250}$. It has to be mention that x, from SiO_x, defines the material composition being the ratio between the content of oxygen and silicon atoms from the deposited layer. By definition, x_{plasma} is a ratio of the light intensity emitted by the excited oxygen and silicon atoms in plasma volume. With a calibration curve of x from the SiO_x composition versus x_{plasma}, *the composition of the deposited layer can be direct and in real time monitored.* This is the advantage of using the OES systems on the deposition equipments based on plasma. Such an example is shown in Figure 6. The values for the x parameter that defines the layer composition have been determined via RBS measurements. This technique determines directly the number of atoms of different species.

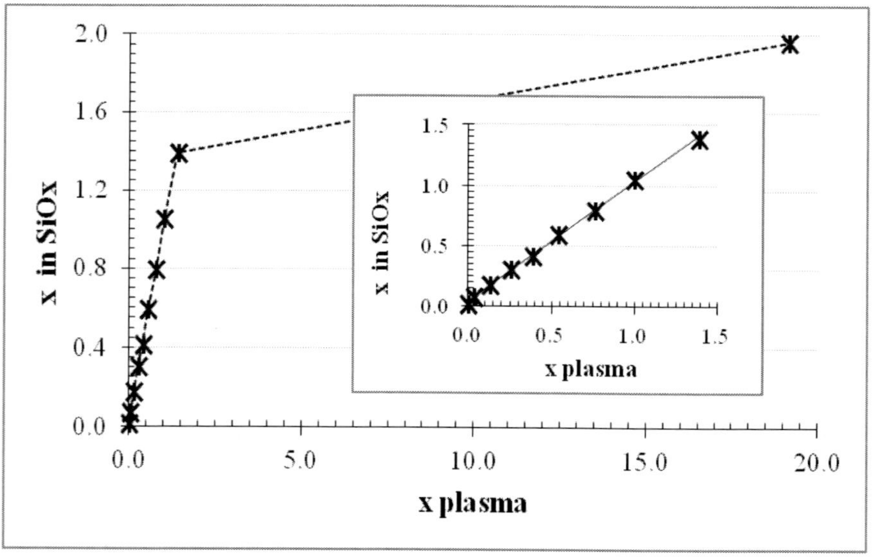

Figure 6. The composition in the deposited SiO_x layer versus the ratio of the OES representative lines from the plasma spectra ($x_{plasma} = \Im_{O\_777} / \Im_{Si\_250}$).

The plots from Figure 6 reveal linearity between the composition in the solid phase (SiO_x thin film) and that in the plasma, when the deposition is not in the so-called "oxidic" phase (x<1.5). The last experimental point represents $SiO_{1.96}$ and was obtained with a plasma where $x_{plasma} = \Im_{O\_777} / \Im_{Si\_250} = 19.13$. When the content of oxygen in the deposition vessel is so high that the target oxidation occurs during the sputtering, the sputtered material is a mixture of silicon and oxygen atoms and not only silicon. This means decreasing the silicon production and an increase in the oxygen content in the plasma phase. The process is function of the sputtering rate and oxidation dynamics.

Before we discuss the stability of the OES signal, we mention that the calibration curve x_SiO_x versus x_{plasma} is dependent on the gas pressure in the vessel and on the rf power used for deposition. Therefore the composition of the deposited SiO_x layer can be tuned in real time during deposition for a deposition system where the reproducibility of plasma versus the deposition conditions is well known.

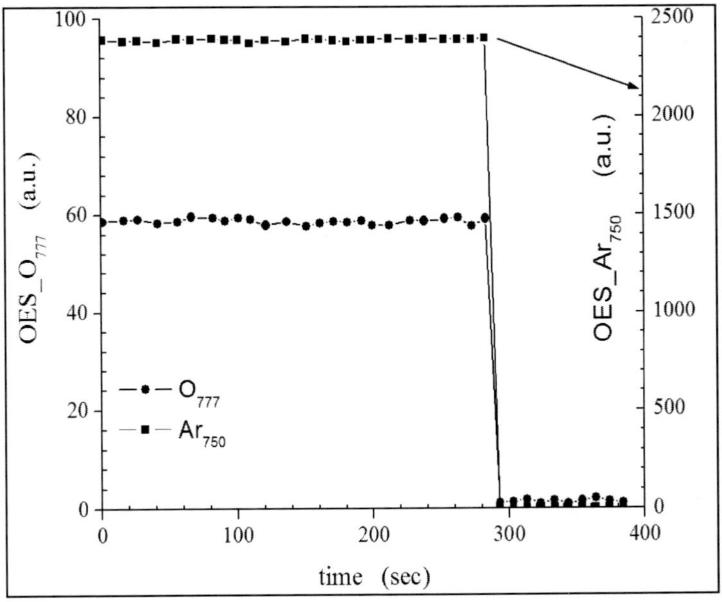

Figure 7. The OES signals for the representative wavelength of Ar and O emissions in plasma versus time; the reliability in the intensity level reveals the plasma stability.

The stability of plasma can be revealed by OES measurement if the data collection is made at different moments of time keeping unchanged the plasma conditions. Such a result is shown in Figure 7 for the signals Ar_750 and O_777 measured simultaneously in the same plasma. We note the difference between the levels of the two collected signals but mainly the steadiness of these levels in time. The latter reveals the stability of plasma during the deposition. This is an important parameter of the process that the user of the deposition system wants to monitor and to control. With electronics that use the OES signal like input, he can act either on the rf power or on the gas flow in order to tune the plasma properties in such a way that the composition of the deposited material is the desired one. In other words, the users can build-up a plasma controller with this OES monitoring system.

4.2. OES on Plasmas Used to Deposit a-C:H Thin Films by PECVD

4.2.1. Introductive Notions on the PECVD Deposition of a-C:H

PECVD systems are successfully used to deposit hydrogenated amorphous carbon (a-C:H) by decomposition of gasses that contain carbon atoms in their molecule: methane CH_4, acetylene C_2H_2, etc. The deposition system contains two electrodes used to "put on" the plasma of a gas (Ar, He or N_2) in mixture with the reactive gas to be decomposed. As usual in this type of deposition systems, one electrode is rf powered (small in dimension) and the other one is grounded. The substrate, where the thin film is grown, is placed on the non-grounded electrode. In the plasma volume, due to collisions with electrons, neutrals and/or ions, the molecules of the reactive gas are dissociated. Supposing that the reactive gas is CH_4, the result of this dissociation can be either a free carbon atom or its combinations with hydrogen atoms produced by dissociation in plasma, too.

Before we get into details regarding the OES plasma analysis, some experimental details are to be given. The rf plasmas have been generated with a 13.56 MHz rf power supply applied to capacitively coupled parallel electrodes. Depositions of a-C:H from CH_4 using plasmas of methane with /without Ar or He or N_2 have been investigated considering the gas flow rates as 5sccm/45sccm for CH_4-Ar and CH_4-He, and 5sccm/0.5sccm for CH_4-N_2. Films were grown at a fixed pressure (100 mTorr) but with different rf powers (i.e., 30W, 120W and 210W). As a result, different negative dc bias voltages on the electrode where the substrates have been mounted were developed. The generated electric field in this plasma sheath accelerates ions towards the substrate and the substrate temperature can increase during deposition. Also the structure and the composition of the deposited material are influenced by this bombardment.

Optical emission spectroscopy (OES) study of the plasma discharge was performed in the range 300-900 nm, through an optical fiber fixed on a quartz window mounted in the chamber-reactor wall.

4.2.2. OES Spectra of Various CH_4 Plasmas: The Main Lines

Typical emission spectra of rf plasmas from CH_4, CH_4-He, CH_4-Ar and CH_4-N_2 mixtures are shown in Figure 8 (a, b, c and d, respectively). The most intense lines in the CH_4 spectrum (Figure 8a) are due to the emissions from CH radicals (390 nm and 430 nm), atomic hydrogen (H_α at 656 nm, H_β at 486 nm and H_γ at 434 nm) and molecular hydrogen (H_2 bands centered at 463 nm and 602 nm) [31].

When together with methane another gas is used, the emission spectra of these plasmas (CH_4-He, CH_4-Ar and CH_4-N_2 in Figure 8) are a combination of the already discussed CH_4 OES lines and Ar, He lines or N_2 bands. As it is shown in Figure 8b-d, the representative lines/bands of the additive gas transitions are more significant in OES spectra. Interesting observations on these OES spectra are briefly hereunder discussed:

i) the OES spectrum of the CH_4-He plasma (Figure 8b) shows a system of peaks due to atomic and ionic He transitions [32]. Considering their intensity, we mention the peaks centered at 389, 447, 502, 588, 668 and 706 nm wavelength;
ii) the OES spectrum of the CH_4-Ar plasma, (Figure 8c) shows a system of emission lines assigned to Ar: the most intense lines at 750 nm and 811 nm form Ar I group, while the group placed in the 415 nm - 490 nm region are ascribed to the Ar II lines;
iii) the optical emission spectrum of the CH_4-N_2 plasma (Figure 8d) reveals peaks assigned to molecular nitrogen: at 316, 337, 380, 400 and 420 nm, are due to the second positive molecular bands of N_2, lines at 391 and 428 nm are due to the first negative system of N_2^+, while peaks centered at 540, 580, 650 and 750 nm are associated with the first positive system of the N_2. In this spectrum, the band at 388 nm is considered to be a signature of the CN molecules formed in the gas phase from excited molecules of CH_x and N_2.

These studies of different additive gases in plasma are aimed to tune the deposition conditions in order to obtain a-C:H layers with desired physical properties. For example the structure of the deposited carbon can be tuned more diamond like (DLC) or graphitic like; the content of nitrogen, when CH_4-N_2 plasmas are used for deposition has a great impact on the electrical conductivity of the a-C:H layers. The optical properties (transparency, refractive index, band gap) are dependent on the structure and composition of the deposited material. As it was discussed in the previous section, with a good understanding of the plasma properties revealed by the OES spectra and their correlation to the physical characteristics of the obtained a-C:H the deposition can be tuned in real time.

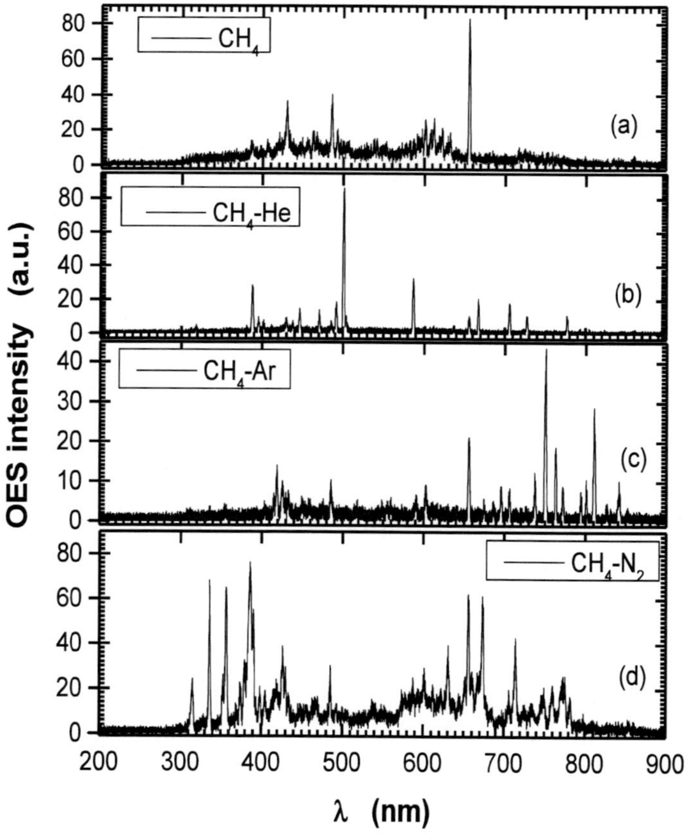

Figure 8. Typical optical emission spectrum from (a) CH_4, (b) CH_4-He, (c) CH_4-Ar, and (d) CH_4-N_2 rf PECVD plasmas.

From the spectra presented in Figure 8 the CH line and the series of H lines are to be analyzed. The main mechanisms to produce these lines are hereunder described:

$$\begin{aligned} CH_4 + e &\rightarrow CH + H + H_2 + e \\ H_2 + e &\rightarrow 2H + e \\ H + e &\rightarrow H^+ + 2e \end{aligned} \qquad (15)$$

where the maximum cross-section for the first reaction is $0.313x10^{-16}$ cm^2 and for the second one (it describes the dissociation of the molecular hydrogen) is $0.898x10^{-16}$ cm^2 [33]. The ionization of the hydrogen atom is characterized by a cross-section of $6.88x10^{-16}$ cm^2 and needs higher energies for the incident electron.

Knowing the mechanisms that govern the optical emissions, it is instructive to see the influence of the nature of gases from plasma and how the rf power "controls" these mechanisms. The intensity values of the CH_430 and the H_α lines versus the used rf power are given in Figure 9. Different plasmas based on CH_4 (non)diluted in He, Ar or N_2 have revealed a linear increase of the CH line intensity with the rf power. In the case of the H_α line (λ=656nm), the intensity increases for all mixtures. When comparing the CH intensity values, having as reference the values obtained in CH_4 plasma, we can see that: a) He increases the CH intensity, b) Ar slowly increases except for the highest rf power investigated, c) N_2 increases the CH intensity. A similar comparison for the H_α line intensity reveals: a) He reduces the line intensity, b) Ar promotes the increase of the H_α intensity, c) N_2 does not produce noticeable changes. At this point, we note the small amount of the used N_2 in comparison with Ar or He in the gas mixture with CH_4.

Most of the investigated lines are obtained by the atom excitation as a result of a collision with one electron; mathematically, the intensity of each line is described by relation (10) where the atom nature is specific for each studied case. When the rf power increases, the self-bias voltage at the plasma sheath increases and the energy of electrons from plasma increases. In other words the electron temperature, T_e, increases with rf power. Because the excitation efficiency is a function of the density and energy distribution of the electrons in plasma, the relation between the line intensity and the applied rf power is direct. For this, let's consider one representative line for each atomic species and follow its intensity variation when the rf power increases. The experimental data for Ar_750, He_501 and N_2_358 are given in Figure 10. The CN presence in plasma is also important because nitrogen in small amounts in the structure of the a-C:H has a determinant effect on electric properties of the deposited layer. Therefore, CN_388 line is monitored, too.

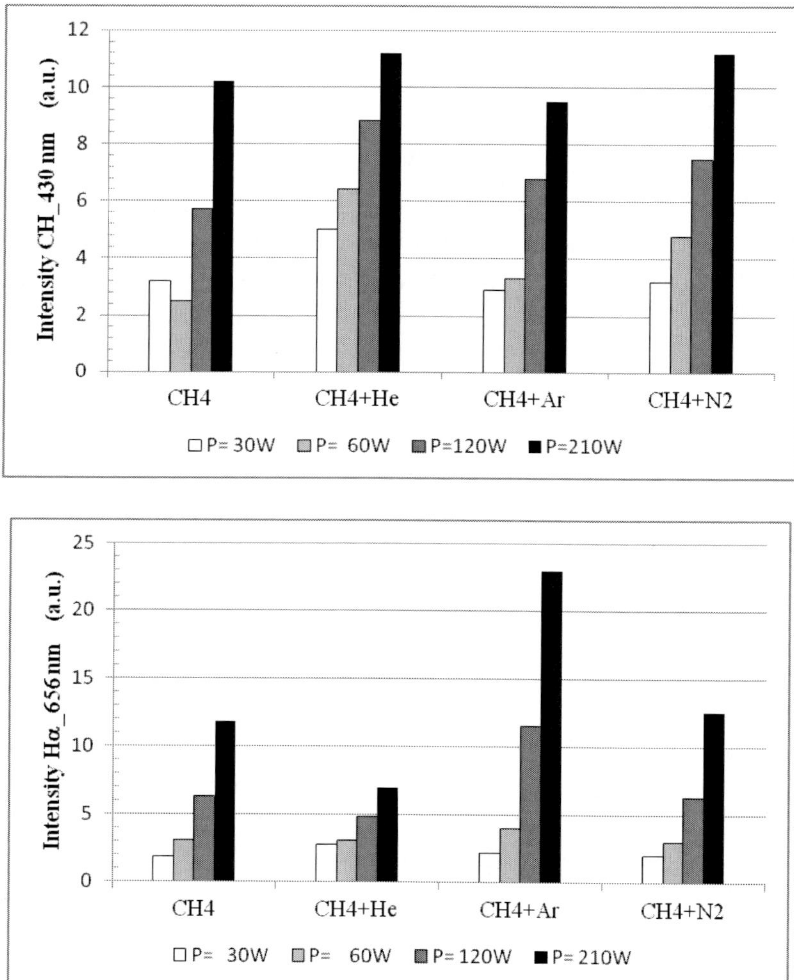

Figure 9. The OES intensity values for the CH and H_α lines (λ=430 nm and λ =656 nm respectively) measured in plasmas of CH_4, CH_4+He, CH_4+Ar, and CH_4+N_2, obtained with different rf power values.

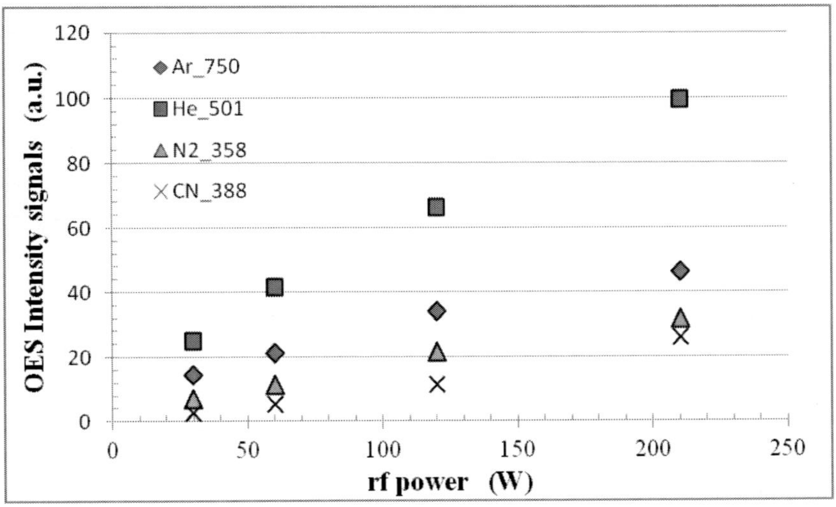

Figure 10. The representative lines for various added gases in plasmas (see the text) versus the rf power. More power means larger amount of collisions and from here more produced light due to de-excitations. We note the parallel shapes of the plots N_2_358 =f(rf power) and CN_388=f(rf power).

Although He and Ar are in the same proportion in plasmas with CH_4 (45 sccm versus 5 sccm of CH_4) their increase in intensity is different when the rf power increases. This may be related to the efficiency of the excitation process and the electron energy distribution. Multiple activation pathways could be also invoked to explain these differences (the ratio of the lines intensities He_501:Ar_750 is 1:2). An extensive discussion on the excitation mechanisms of Ar atoms/ions in plasma has been presented in section 4.1. Specific information about the He signal is given hereunder:

- In a CH_4-He plasma, excited He (2^3S), He (2^1S) and He^+ ($2S$) species are generated, which have reaction rate coefficients with CH_4 molecules of 2, 4 and 13 $\times10^{-10}$ cm^3s^{-1}, respectively [34]. These species can generate CH_n^+ ion via Penning ionization and dissociation. In particular, with an excitation energy of 19.8 eV (5 eV below the ionization potential of He), the He (2^3S) atoms can play a major role in producing CH_4^+ and CH_3^+ ions.
- In the CH_4-He plasmas the He^+ and/or the hydrocarbon ions contribute to the maintenance of the plasma. The ionization potential of He (24.5 eV) is much higher than the threshold potentials for the

generation of CH_4^+ and CH_3^+ ions from CH_4 molecules (12.6eV and 14.3eV, respectively), thus at low rf power, Penning ionization will not be significant. Increasing the rf power, the increased densities of these metastable He atoms lead to an increase of the CH_4^+ and CH_3^+ ions density; this, in turn, alters the balance of the CH_4 fractionation reactions and may account for the increased CH emission and reduced H intensities observed in the CH_4-He plasmas relative to CH_4 (Figure 9).

The amount of nitrogen in N_2+CH_4 plasmas is much smaller than the helium amount in He+CH_4 plasmas (N_2:He is 1.1 :100); this proportion is not reflected in the plot from Figure 10 where the ratio of the lines intensities N_2_358:He_501 is 1:3. These examples show that the processes in plasma are complicated and the OES is a good tool to investigate them.

4.2.3. Ideas for Process Characterization and Method Calibration

The first aspect checked during or after a thin film deposition is the growth rate. This is very important from a deposition efficiency point of view and, why not, for the costs involved in obtaining these layers. For the PECVD deposited a-C:H thin films discussed in this work, the deposition rates as a function of rf power, are shown in Figure 11. Increasing the rf power values, the deposition rates of films grown from CH_4/He plasmas, although lower, follow the same trend as those deposited from pure CH_4 plasmas. For depositions from CH_4 diluted in Ar or with N_2 addition, the deposition rate trends are quite different when comparing with the depositions from pure CH_4. At low rf power, N_2 has little effect on the deposition rate but there is a significant linear reduction with the increasing rf power. With Ar dilution, at low rf power, the deposition rate tracks that of the He dilution but increasing the rf power, it falls off to the level observed for the N_2 diluted films. The decrease in the film growth rate when rf power is increased is due to the energetic ion bombardment of the just deposited film surface. The electric field developed over the plasma sheath region is responsible for this bombardment and its efficiency.

The deposition rate is a balance between the deposition on the substrate surface and the sputtering of the deposited atoms / entities due to this ion bombardment. The interaction with ions has also benefic results: the composition can be tuned (e.g., the bombardment with Ar ions results in a film of a-C:H with a lower hydrogen content [35]) and the compactness of the deposited film is improved.

More efficient is the sputtering in the plasmas that contain N_2^+ ions. This species acts as an efficient chemical/physical sputtering agent despite the low concentration of N_2 (N_2/ CH_4 ~10%). Etch rate measurements show a sputter yield for N_2 with 2 to 3 orders of magnitude larger than that of Ar in the range of rf power here used [15].

The properties of the deposited material are strongly affected by the deposition conditions. In Figure 12a the optical band gap – a well known parameter in physics and electronics - is represented versus the rf power used for a-C:H deposition from CH_4+He plasmas. The higher is the rf power, the lower is the value of the optical band gap. Of course, the next step is to relate these values to the measured signals via OES; and this is shown in Figure 12b. The correlation between the OES parameter measured during the deposition and the optical band gap measured post-deposition on the solid thin films is evident. With adequate electronics and programming the deposition can be tuned direct and in real time in order to obtain material with desired properties. The Optical Emission Spectroscopy (OES) technique is a versatile and useful tool in plasma depositions of thin solid films.

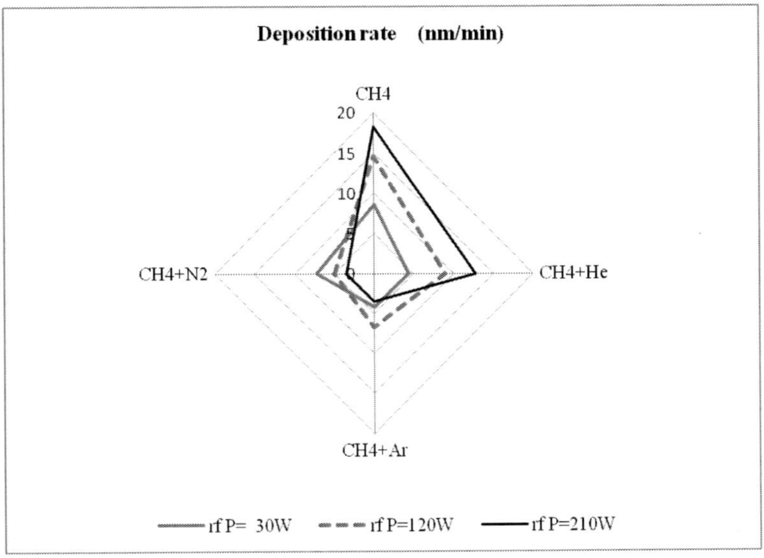

Figure 11. The deposition rate for a-C:H deposited from different plasmas (CH_4, CH_4+He, CH_4+Ar, and CH_4+N_2) obtained with different rf power values: 30W, 120W and 210W.

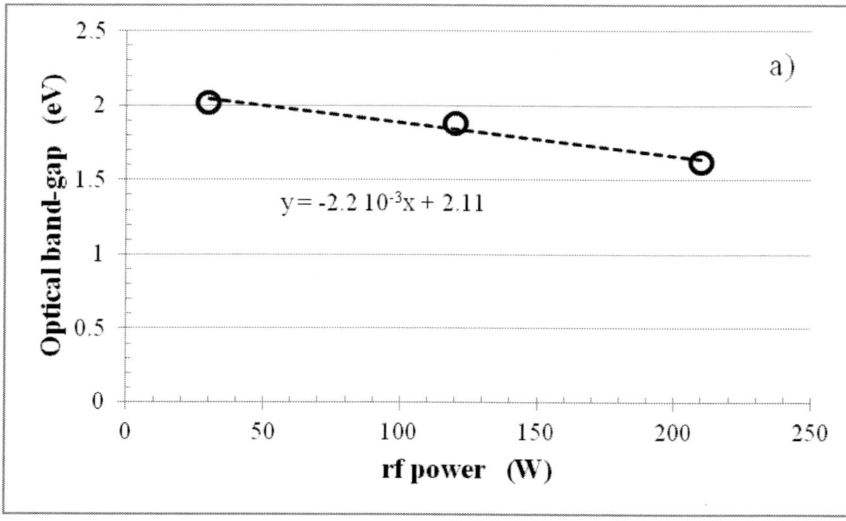

(a)

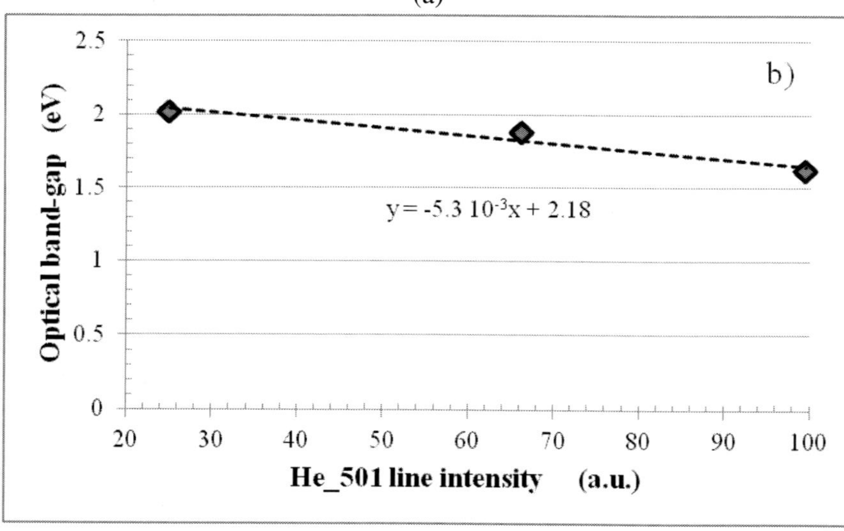

(b)

Figure 12. Optical band gap for samples of a-C:H deposited with CH_4+He plasmas as a function of the rf power (a): the higher the used rf power, the smaller is the band gap of the obtained material. The same optical band gap data represented versus the OES intensity of the He_501 line (b).

CONCLUSION

The aim of this work is to introduce to the reader the optical emission spectroscopy as a tool to i) investigate plasmas used in depositions of thin layers of semiconductors/dielectrics, b) foresee the properties of the deposited materials from the gas phase, and c) tune direct and in real time the deposition conditions in order to obtain the material with desired properties.

The main processes in argon plasmas used either to sputter silicon atoms from a polycrystalline target or to decompose a gas carrier such as CH_4 for carbon atoms have been presented to understand the involved mechanisms in producing the plasma light and the deposition of thin film. Mechanisms as atom excitation / ionization as a result of collisions with energetic electrons, neutrals or ions from the plasma volume have been presented and discussed. The atoms actively involved in the composition/structure of the deposited layer have their own signatures in the plasma light; we have shown that, by investigating the optical lines from the plasma spectrum, knowledge related to the deposition process can be build-up and important correlations to the physical properties of deposited material can be established.

The method of optical emission spectroscopy was described in general terms and then particularized on two processes: i) the obtaining of silicon suboxide SiO_x ($0<x<2$) by rf reactive magnetron sputtering, and ii) the deposition of hydrogenated amorphous carbon (a-C:H) by PECVD. In both cases detailed examples of the influences of the plasma properties on the characteristics of the deposited material (composition, structure, optical properties, electrical properties, etc.) are given. The plasma properties have been revealed by the OES spectra and presented in the light of the mechanisms involved in plasmas. Parameters like rf power, gas pressure and gas composition/nature have been modified during plasma deposition and their influence on the OES spectra have been investigated. Moreover, the influence of these parameters on the deposition rate of the solid material was shown and explained. Also, a connection to the physical properties of SiO_x and a-C:H layers has been made.

Through this work we hope that we could reveal the advantages of using the Optical Emission Spectroscopy (OES) technique as a tool in plasma depositions of thin solid films to diagnose plasma and to tune in real time the deposition to obtain the layers with desired properties. In our examples here presented it worked very well for both, the research investigations and industrial applications.

REFERENCES

[1] N. Mizutani, N. Wakiya, M. Yoshida, K. Hijikata and K. Shinozaki, "Preparation of epitaxial YSZ thin film on Si(001) using metal and oxide targets by RF- magnetron sputtering," *Ferroelectrics,* vol. 260, pp. 249-254, 2001.

[2] R. Stauber, J. Perkins, P. Parrilla and D. Ginley, "Thin Film Growth of Transparent p-type CuAlO2," *Electrochem. Solid State Lett.,* vol. 2, pp. 654-656, 1999.

[3] D. Dobkin and M. Zuraw, Principles of Chemical Vapor Deposition, Kluwer, 2003.

[4] R. Schropp, B. Stannowski, A. Brockhoff, P. v. Veenendaal and J. Rath, "Hot wire CVD of heterogeneous and polycrystalline silicon semiconducting thin films for application in thin film transistors and solar cells," *Materials Physics and Mechanics,* vol. 1, pp. 73 - 82, 2000.

[5] A. Rahtu, Atomic Layer Deposition of High Permittivity Oxides: Film Growth and In Situ Studies (Thesis), Helsink: University of Helsink, 2002.

[6] A. Stick, M. Markmann, K. Brunner, G. Abstreiter and E. Muller, "Silicon/silicon suboxide heterostructures grown by molecular beam epitaxy," *Materials Science and Engineering,* vol. 89, no. 1-3, pp. 274-278, 2002.

[7] M. Zacharias, T. Drusedau, A. Panckow, H. Freistedt and B. Garke, "Physical properties of a-SiOx:H alloys prepared by dc magnetron sputtering with water vapour as oxygen source," *J. Non-Cryst. Solids,* vol. 169, no. 1, pp. 29-36, 1994.

[8] G. Lucovsky, "Atomic structure and thermal stability of silicon oxides thin films and transition regions at Si-SiO2 interfaces," *J. Non-Cryst. Solids,* Vols. 227-230, no. 1, pp. 1-14, 1998.

[9] N. Tomozeiu, E. v. Fassen, W. Arnoldbik, A. Vredenberg and F. Habraken, "Structure of sputtered silicon suboxide single- and multi-layers," *Thin Solid Films,* Vols. 421-422, p. 382 – 385, 2002.

[10] A. Barranco, J. Meijas, J. Espinos, A. Caballero, A. Gonzales-Elipe and F. Yubero, "Chemical stability of Si_n+ species in SiOx (x<2) thin films," *J. Vac. Sci. Technol.,* vol. 19, no. 1, pp. 136-144, 2001.

[11] J. v. Hapert, Hopping Conduction and Chemical Structure a study on silicon suboxides, Utrecht: Utrecht University PhD Thesis, 2002.

[12] H. Tsai and D. Bogy, "Characterization of diamondlike carbon films and their application as overcoats on thin-film media for magnetic recording," *J. Vac. Sci. Technol., A* , vol. 5, pp. 3287-3312, 1987.

[13] J. Angus and C. Hayman, "Low-Pressure, Metastable Growth of Diamond and "Diamondlike" Phases," *Science,* vol. 241, pp. 913-921, 1988.

[14] A. Bogus, I. Gebeshuber, A. Pauschitz, R. Manish and R. Haubner, "Micro- and nanomechanical properties of diamond film with various," *Diamond & Related Materials,* pp. 1998 - 2004, 2008.

[15] K. Clay, S. Speakman, N. Morrison, N. Tomozeiu, W. Milne and A. Kapoor, "Material properties and tribological performance of rf-PECVD deposited DLC coatings," *Diamond & Related Materials,* vol. 7, pp. 1100 - 1107, 1998.

[16] S. Lansley, H. Looi, M. Witfield and R. Jackman, "An optically activated diamond field effect transistor," *Diamond & Related Materials,* vol. 8, pp. 946 - 951, 1999.

[17] P. Bergonzo, A. Brambilla, D. Tromson, C. Mer, B. Guizard, R. Marshall and F. Foulon, "CVD diamond for nuclear detection applications," *Nuclear Instruments and Methods in Physics Research,* vol. 476, no. 3, pp. 694 - 700, 2002.

[18] N. Tomozeiu and W. Milne, "Magnetic multipole confinement used in amorphous hydrogenated carbon films deposition," *Journal of Non-Crystalline Solids,* vol. 249, pp. 180 - 188, 1999.

[19] A. Bogaerts and R. Gijbels, "Comparison of argon and neon as discharge gases in a direct-current glow discharge a mathematical simulation," *Spectrochimica Acta, Part B,* vol. 52, no. 5, pp. 553 -565, 1997.

[20] E. Eggarter, " Comprehensive optical and collision data for radiation action," *J. Chem. Phys.,* vol. 62, no. 3, pp. 833-847, 1975.

[21] A. Bogaerts and R. Gijbels, "The role of fast argon ions and atoms in the ionization of argon in a direct-current glow discharge: A mathematical simulation," *J.Appl. Physics,* vol. 78, pp. 6427-6431, 1995.

[22] N. Fuller, M. Malyshev, V. Donnelly and I. Herman, "Characterization of transformer coupled oxygen plasmas by trace rare gases-optical emission spectroscopy and Langmuir probe analysis," *Plasma Sources Sci. Technol.,* vol. 9, no. 2, pp. 116-127, 2000.

[23] T. Ombrello, S. Won, Y. Ju and S. Williams, "Flame propagation enhancement by plasma excitation of oxygen," *Combustion and Flame,* vol. 157, pp. 1916 - 1928, 2010.

[24] A. Phelps, "ftp://jila.colorado.edu/collision_data/electron.txt," 2009. [Online].

[25] E. v. Hattum, RF reactive magnetron sputter deposition of silicon sub oxides, Utrecht University, PhD Thesis, 2007.

[26] N. Tomozeiu, "On the sputtering deposition process of silicon suboxide thin films," *J. of Optoelectronics and Advanced Materials,* vol. 4, no. 4, pp. 959-964, 2002.

[27] N. Tomozeiu, "Silicon Oxide (SiOx, 0<x<2): a challenging material for optoelectronics," in *Optoelectronics - Materials and Techniques,* InTech, 2011, pp. 55-98.

[28] http://physics.nist.gov, "NIST Home/ PML / Physical Reference Data / Atomic Spectra Database," December 2013. [Online].

[29] R. Walkup, K. Saenger and G. Selwyn, "Studies of atomic oxygen in O2+CF4 rf discharges by two-photon laser-induced fluorescence and optical emission spectroscopy," *J. Chem. Phys.,* vol. 84, no. 5, pp. 2668-2674, 1986.

[30] N. Tomozeiu, "Optical emission studies of the gas phase during the silicon suboxide deposition by reactive rf magnetron sputtering," *Thin Solid Films,* vol. 515, pp. 6582-6585, 2007.

[31] C. Barholm-Hansen, M. Bentzon and J. Hansen, "Optical emission spectroscopy during growth of diamond-like carbon from a methane plasma," *Diamond and Related Materials,* vol. 3, no. 4-6, pp. 564-568, 1994.

[32] E. d. l. Cal, D. Taffala and F. Tabares, "Characterization of He/CH4 dc glow discharge plasmas by optical emission spectroscopy, mass spectrometry, and actinometry," *J. Appl. Phys., 73, (1993) 948.,* vol. 73, no. 2, pp. 948-954, 1993.

[33] H. Tahara, K. Minami, A. Murai, T. Yasui and T. Yoshikawa, "Diagnostic Experiment and Kinetic Model Analysis of Microwave CH4/H2 Plasmas for Deposition of Diamondlike Carbon Films," *Jpn.J. Appl. Phys. 34 (1995) 1972.,* vol. 34, pp. 1972-1979, 1995.

[34] N. Mutsukura and K. Miyatani, "Deposition of diamond-like carbon film in CH4/He r.f. plasma," *Diamond and related Materials,* vol. 4, pp. 342-345, 1995.

[35] T. DebRoy, S. Kumar and K. Tankala, "Optical emissions during plasma assisted chemical vapor deposition of diamond-like carbon films," *Diamond and Related Materials, 4 (1994) 69.,* vol. 4, no. 1, pp. 69-75, 1994.

[36] L. Martinu, O. Zabeida and J. Klemberg - Sapieha, "Plasma_Enhanced Chemical Vapor Deposition of Functional Coatings," in *Handbook on Thin Film Deposition Technologies*, Amsterdam , Elsevier, 2009, pp. 394 - 467.

In: Optical Spectroscopy
Editor: Nicolae Tomozeiu

ISBN: 978-1-63321-197-1

Chapter 2

REFLECTANCE SPECTROSCOPY

María Gabriela Lagorio*

INQUIMAE/ Dpto. de Química Inorgánica, Analítica y Química Física, Facultad de Ciencias Exactas y Naturales. Universidad de Buenos Aires, Ciudad Universitaria, Buenos Aires, Argentina

ABSTRACT

The description, analysis and applications of Reflectance Spectroscopy in ultraviolet, visible and infra-red regions of the electromagnetic spectrum are reviewed. A comprehensive introduction is given to address the theoretical basis for the analysis of reflected light from scattering materials. Differences between specular and diffuse components are presented and discussed in terms of particle size. The Kubelka Munk theory, the Remission Function and the Pile of Plates model are also introduced and the calculation of absorption and scattering factors is presented. The obtention of chromophore concentrations from reflected light is also shown and several examples are given. Steady–state Reflectance, FTIR Reflection techniques and Time Resolved Reflectance spectroscopies are discussed and the equipments are described in detail. Applications in remote sensing, in plant physiology, in soils science, in diseases diagnostics as liver pathology and neonatal jaundice, in detection of tissue oxygenation, in ecological studies for prediction of the

* E-mail: mgl@qi.fcen.uba.ar.

composition of plants and animal tissues, in the study of cultural heritage objects, among several others are also reviewed.

Keywords: Reflectance, Absorption, Attenuated total reflectance, Scattering, Remote sensing

1. Absorption, Transmission and Reflection of Light. Specular and Diffuse Reflection

In the most general case, when a light beam reaches a given surface, part of the light is absorbed, part is transmitted and a portion is reflected.

The light reflected consists of two parts: the regular reflection (also called mirror or specular) and the diffuse reflection.

The specular reflection (R_s) follows the Fresnel equations and for perpendicular incidence, the ratio between the intensities for the reflected (I) and the incident beam (I_0) may be written as (Wendlandt, 1966):

$$R_S = \frac{I}{I_0} = \frac{(n-1)^2 + n^2\kappa^2}{(n+1)^2 + n^2\kappa^2} \tag{1}$$

where n is the refractive index for the irradiated layer of material; and κ is its absorption index defined by the Lambert law (equation 2):

$$I = I_0 \exp\left[-4\pi n\kappa d / \lambda_0\right] \tag{2}$$

In equation (2), d is the layer thickness and λ_0 is the wavelength of the radiation in vacuum (Wendlandt, 1966). The diffuse reflection, on the other hand, is connected with the penetration of a portion of the incident flux into the interior of the illuminated layer. A part of this radiation is absorbed and the rest is returned to the surface after multiple scattering at the boundaries of small particles in the sample. Mirror reflectivity prevails in smooth surfaces like mirrors. In microscopically rough surfaces (clothes, walls, etc.) the beam reflects in different directions given place to diffuse reflectance (Figure 1).

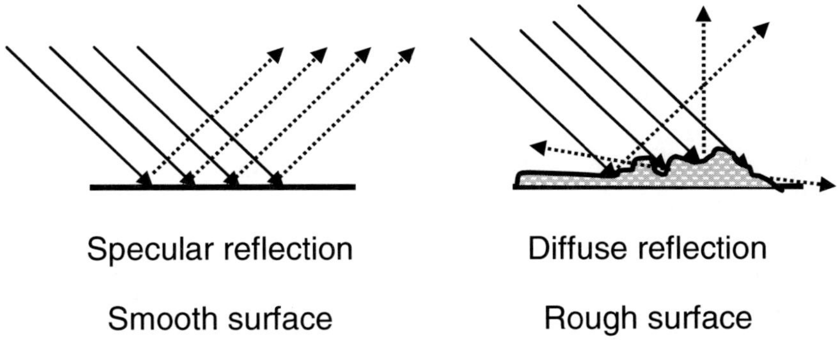

Figure 1. Schematic description of specular and diffuse reflections.

In Figure 1, it can be observed that the general law of reflection, predicting that the light ray reflects with an angle of reflection equals to the angle of incidence, is fulfilled both in the case of a smooth surface as in a rough surface. However, in the second case, the net result is a collection of beams directed towards different angles in the space.

When the angular distribution of the reflected light is independent of the angle of incidence, the diffuse reflection is defined as "ideal diffuse" (Kortum, 1969). Lambert proposed the first law of diffuse reflection based on the observation that a white wall illuminated by sunlight is seen equally bright at all the observation angles (Lambert, 1760).

The attenuation of the diffuse component of the reflected light by absorption within the object is described by equation (3) known as the Bouguer- Lambert Law:

$$I = I_0 \exp\left[-\varepsilon' \bar{d}\right] \tag{3}$$

where ε' is the chromophore molar extinction coefficient and $\bar{d}$ is the mean penetrated layer thickness (Wendlandt, 1966).

The ratio $R= I/I_0$ $(0 < R < 1)$ is defined as the *reflecting power*. The *percentage reflecting power* is expressed as *100.R* and it is also frequently used. When describing the reflecting power associated with the specular reflection the term *reflectivity* is employed while for diffuse reflection the word *reflectance* is preferred. When both components are present it is frequent to refer to the reflection processes as specular reflectance and diffuse reflectance respectively.

2. Kubelka Munk Theory. The Remission Function

For a quantitative description of the interaction of light with diffusing media, the Kubelka–Munk theory is usually applied. This theory only applies to diffuse reflection. The Kubelka–Munk theory assumes that the medium consists of randomly distributed, uniformly absorbing scattering particles. Two light fluxes, *I* (the flux in the direction of the incident light) and *J* (the flux in the direction of the reflected light), are considered to travel in opposite directions perpendicular to the irradiated surface at x = 0 (Figure 2).

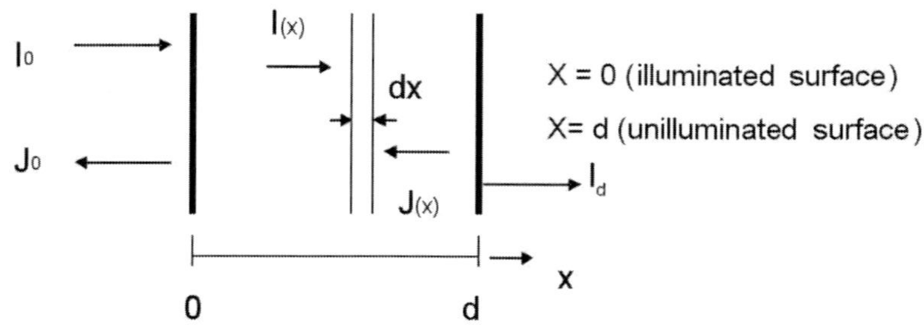

Figure 2. Representation of the two light fluxes considered in the Kubelka Munk theory for a layer of light absorbing and light scattering particles

Travelling through the layer, the intensity of the photon flux decreases. As the absorbed and the scattered part of *I* decreases *I* and the scattered part of *J* increases *I*, the attenuation (*dI*) may be expressed as:

$$dI(x) = -I(x)\,(k+s)\,dx + J(x)\,s\,dx \tag{4}$$

where *k* and *s* are the absorption and scattering coefficients respectively: $k = 2\varepsilon$ and $s=2\sigma$, with ε the fraction of light absorbed and σ the fraction of light scattered, respectively, per unit path length in the sample (Wendlandt, 1966).

For *J*, the originated light flux travelling in the opposite direction to *I*, a similar equation may be written:

$$dJ(x) = J(x)\,(k+s)\,dx - I(x)\,s\,dx \tag{5}$$

For *J*, conversely, the absorbed and the scattered part of *J* decreases *J*, and the scattered part of *I* increases *J* (see equation 5). It should be noted that the signs in equations 4 and 5 have been chosen consistent with the fact that as *x* increases *I* decreases, while *J* increases.

Equation 5 is solved for J and substituted in equation 4 to give:

$$\frac{d^2I}{dx^2} - \kappa^2 I = 0 \tag{6}$$

Similarly,

$$\frac{d^2J}{dx^2} - \kappa^2 J = 0 \tag{7}$$

The parameter κ is related to the absorption and scattering coefficients according to:

$$\kappa^2 = k(k + 2s) \tag{8}$$

Solving equations (6) and (7) taking into account the boundary conditions: $I = I_0$ for $x = 0$ and $J = 0$ for $x = d$, allows calculation of the transmittance (T) and diffuse reflectance (R) of the considered layer:

$$T = \frac{I_d}{I_0} = \frac{4\beta}{(1+\beta)^2 e^{\kappa d} - (1-\beta)^2 e^{-\kappa d}} \tag{9}$$

$$R = \frac{J_0}{I_0} = \frac{(1-\beta^2)\left(e^{\kappa d} - e^{-\kappa d}\right)}{(1+\beta)^2 e^{\kappa d} - (1-\beta)^2 e^{-\kappa d}} \tag{10}$$

where β is:

$$\beta = \frac{\kappa}{k + 2s} \tag{11}$$

When $d \rightarrow \infty$ (infinite layer), $T \rightarrow 0$ and

$$R_\infty = \frac{1-\beta}{1+\beta} \tag{12}$$

and

$$\frac{(1-R_\infty)^2}{2R_\infty} = \frac{k}{s} \tag{13}$$

The left side of equation 13 is known as the remission function, *F(R),* and for a constant scattering coefficient, it is linearly dependent on *k* and consequently on the number of absorbing chromophores in any sample (i.e., on the concentration of the absorbing species). From equation 13, it is clear that measuring the diffuse reflectance for an infinite layer (i.e., a layer displaying zero light transmittance) the ratio between the absorption and scattering coefficients (the remission function) may be calculated.

For a mixture of chromophores, the absorption and scattering coefficients may be considered as additive functions of the respective coefficients of the mixture components weighed according to their mass fractions (equation 14). (Duncan 1940 and Duncan 1949)

$$\frac{(1-R_\infty)^2}{2R_\infty} = \frac{k}{s} = \frac{\sum_i c_i.k_i}{\sum_i c_i.s_i} \tag{14}$$

where c_i is the mass fraction and k_i is the absorption coefficient of component *i*. The Kubelka- Munk theory by itself does not allow the calculation of the separate values of *k* and *s*. To estimate these quantities it is necessary to use other models to describe the light interaction with the system as the Pile of Plates model, which is presented in the next section.

3. Pile of Plates Model

Calculation of the absorption and scattering coefficients from reflectance and transmittance measurements is possible when the Pile of Plates model is taken into account.

This model considers the sample composed of a large number of parallel layers (Wendlandt, 1966), (Cabrera et al., 1996), (Gu et al., 2004).

For a pair of layers (*i, j*), when a beam impinges on the layer *i,* a fraction *Ri* is reflected and a portion *Ti* is transmitted (Figure 3). The transmitted beam *Ti* impinges then on the layer *j,* and there a fraction *TiTj* is transmitted and *TiRj* is reflected.

This fraction reaches layer *i* and a fraction *TiRjTi* emerges out from *i.* A fraction *TiRjRi* is reflected at the lower side of *i.* Continuing the description of the light interaction with the layers *i* and *j* in this way, expressions for the transmittance ($T_{i,j}$) and reflectance ($R_{i,j}$) resulting through or from the two layers are obtained:

$$T_{i,j} = T_i T_j + T_i R_j R_i T_j + ... = T_i T_j (1 + R_i R_j + R_i^2 R_j^2 \quad (15)$$

$$T_{i,j} = \frac{T_i T_j}{1 - R_i R_j} \quad (16)$$

$$R_{i,j} = R_i + T_i R_j T_i + T_i R_j^2 R_i T_i + ... = R_i + T_i^2 R_j (1 + R_i R_j + R_i^2 R_j^2 + ... \quad (17)$$

$$R_{i,j} = R_i + \frac{T_i^2 R_j}{1 - R_i R_j} \quad (18)$$

Infinit geometric series were used in equations 15 and 17.

Stokes (Stokes, 1860), (Allen and Richardson, 1968), deduced that for a sample of thickness *d*, equation 19 may be deduced from equations 16 and 18.

$$\frac{R}{b^{\frac{d}{l}} - b^{\frac{-d}{l}}} = \frac{T}{a - a^{-1}} \quad (19)$$

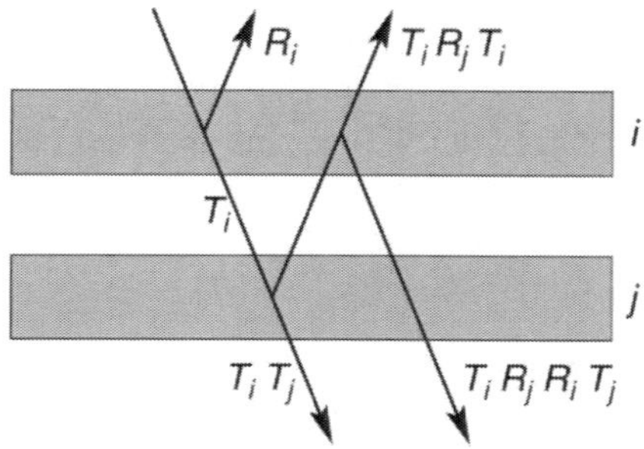

Figure 3. Pile of plates model. Pictorial image for the interaction of light with two plates (i, j). The fraction of reflected light is denoted R and the fraction of transmitted light is denoted T. The subscripts i and j refer to the respective plates. (From Absorption and scattering coefficients. A biophysical-chemistry experiment using reflectance spectroscopy, Gabriela B. Cordon and M. Gabriela Lagorio, *J. Chem. Educ.*, 2007, 84, 1167-1170).

where l is the thickness of each layer in the pile of plates model, R and T are the reflectance and transmittance, and a, b, and Δ are given by equations 19–21 for a system composed of d/l number of layers:

$$a = \frac{1 + r^2 - t^2 + \Delta}{2r} \tag{20}$$

$$b = \frac{1 + t^2 - r^2 + \Delta}{2t} \tag{21}$$

$$\Delta^2 = (1 + r + t)(1 + r - t)(1 + t - r)(1 - r - t) \tag{22}$$

where, r and t refers to the reflectance and transmittance of one layer.

In this way, by considering the pile of plates model, a relation between the Stokes parameters a and b and measurable data (reflectance and transmittance of one layer) is obtained. (Cordon and Lagorio, 2007)

If a comparison among the expressions for T and R (equations 9 and 10) obtained from the Kubelka Munk theory and equation (19) from the Pile of Plates model is performed, equations 23 and 24 can be deduced:

$$\kappa = \log b \tag{23}$$

and

$$\beta = \frac{a-1}{a+1} \tag{24}$$

Equations 23 and 24 assume the simultaneous validity of both the Kubelka Munk theory and the Pile of Plates model for the description of the interaction between light and the layer. Finally, as κ and β are related to k and s (equations 8 and 11), equations 25 and 26 may be written:

$$k = \frac{a-1}{a+1} \log b \tag{25}$$

$$s = \frac{2a}{a^2-1} \log b \tag{26}$$

These equations are very important experimentally because they allow calculation of the absorption and scattering coefficients separately from reflectance and transmittance measurements of a single layer of the studied material.

4. Reflectance and Particle Size

The light reflected by an object and its color depend on the particle size. As an example, the salt $CuSO_4.5H_2O$ appears blue in large crystals and almost white when powdered. On the other hand, some compounds increase their colors upon grinding. This is the case for potassium permanganate, $KMnO_4$, which enhances its color at very small particle sizes. (Lagorio, 2004).

From Fresnel equation (equation 1) it may be noticed that an increase in the absorption index (κ) of the absorbing species, leads to an increment in the observed specular reflectance. On the other hand, decreasing particle size results in a loss of surface smoothness and consequently in a decrease in specular reflection (Wendlandt, 1966), (Körtum, 1969).

For strongly absorbing materials specular reflectance is very important. In fact, from equation 1, it is observed that values for specular reflection approaches 1 when κ tends to infinit. Oppositely, for weak absorbing powders, the specular reflection is smaller.

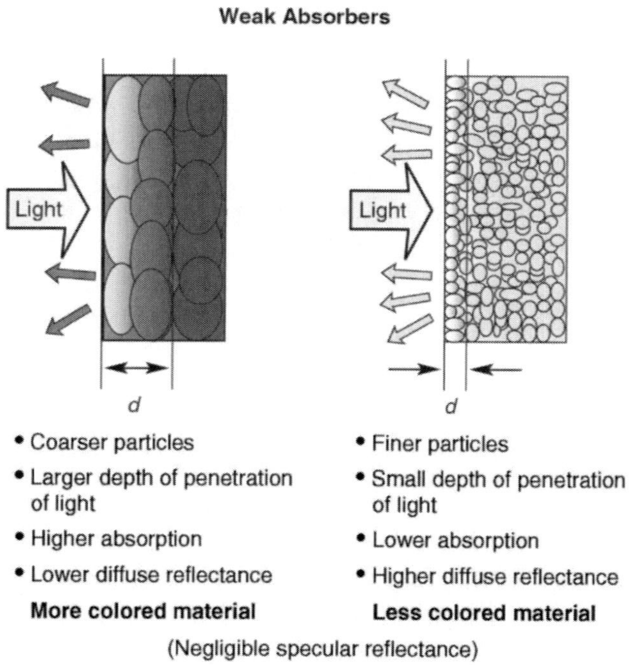

Figure 4a. Pictorial description of the light penetration as a function of the particle size for a weak absorber. Light reflectance is primarily due to diffuse reflection. (From Why do Marbles become paler on grinding? Reflectance Spectroscopy, Color and particle size, M. Gabriela Lagorio. *J. Chem. Educ*., 2004, 81, 1607-1611).

When particle size is decreased, there is an increase in the number of reflections at the grain boundaries and the diffuse reflectance augments. This fact leads to a reduction in the depth of penetration of the light and the absorbed fraction of radiation diminishes. As a result of diminishing particle size, the absorbed part of the radiation is decreased and the reflected portion is augmented. Conversely, with increasing particle size, the radiation can penetrate deeper in the solid, the average path crossed becomes greater and the absorbance increases (Figure 4a). For very strong absorbers, however, the depth of penetration of light is so small that the diffuse reflectance becomes almost independent of particle size. Put another way, changes in particle size for a strongly absorbing material do not result in significant changes in the layer depth penetrated by the light (Figure 4b).

Summarizing, decreasing particle size in a weakly absorbing compound, increases light scattering, decreases light penetration, increases reflectance and consequently decreases light absorption: the color is seen lighter. In a highly absorbing medium, the depth light penetration is so small that it usually becomes independent of particle size. In such case, decreasing particle size, decreases specular reflectance, increases light absorption and the color is seen darker.

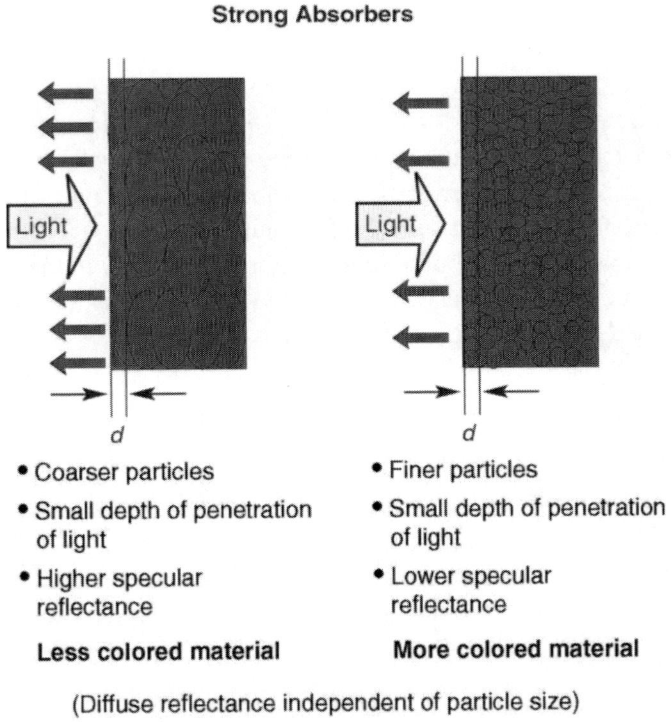

Figure 4b. Pictorial description of the light penetration as a function of the particle size for a strong absorber. Light reflectance is primarily due to specular reflection. (From Why do Marbles become paler on grinding? Reflectance Spectroscopy, Color and particle size, M. Gabriela Lagorio. *J. Chem. Educ.*, **2004**, 81, 1607-1611).

5. MEASUREMENT OF REFLECTANCE

Reflectance measurements can be divided into two general categories: External reflection and Internal reflection.

External reflection is obtained by illuminating an object and analysing directly the beams reflected from the sample surface. Equipments for external reflectance measurements are usually designed to discriminate between specular and diffuse components.

In internal reflection techniques, instead, the interaction of the electromagnetic radiation on the interface between the sample and a medium with higher refractive index is studied.

5.1. UV-visible-NIR Reflection

Reflectance in the UV-VIS-NIR is normally measured with a spectrophotometer equipped with an integrating sphere. Particularly useful are those spectrophotometers covering the electromagnetic spectrum from UV to NIR (usually from 220 to 2600 nm). (Links 1 and 2)

Figure 5 shows a pictorial description of the light beams in the accessory for reflectance measurements used attached to a spectrophotometer. Normally integrating spheres can also measure diffuse transmittance.

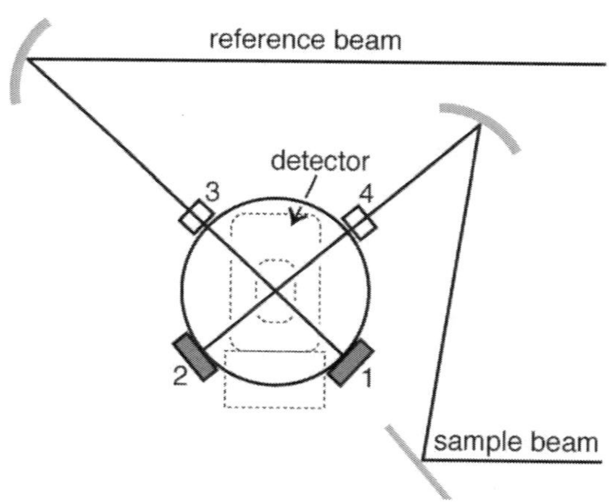

Figure 5. Schematic representation of an integrating sphere. For diffuse reflectance measurements: Barium sulfate is placed in position 1, sample in position 2, air in positions 3 and 4. For diffuse transmittance measurements, Barium sulfate in position 1, air in position 2, air in position 3 and sample in position 4. The detector is at the bottom of the sphere. (From Absorption and scattering coefficients. A biophysical-chemistry experiment using reflectance spectroscopy, Gabriela B. Cordon and M. Gabriela Lagorio, *J. Chem. Educ.*, 2007, 84, 1167-1170).

The 100% reflectance level is adjusted with a Barium sulfate standard. The baseline correction is performed using Barium sulfate in positions 1 and 2 in Figure 5. To record the reflectance spectrum, the reference is placed then at position 1 and the sample at position 2. In some integrating spheres the incidence angle to the reflecting sample may be chosen from two different values. When this angle is set to zero degree, the specularly reflected light is removed from the entrance window and only the diffuse part of the reflectance is recorded. For the other angle choice, total reflectance (specular and diffuse) is obtained.

Rough and irregular surface materials as powders usually present negligible specular reflectance.

5.2. IR Reflection Techniques

In traditional Infra-red spectroscopy, IR radiation is passed through a sample and the fraction of transmitted light is measured as a function of the light wavenumber. IR reflection techniques, instead, arise from the combination of IR spectroscopy with reflection theory. (Khoshhesab, 2012).

In external reflectance, the IR beam is reflected directly from the sample surface.

Internal reflection spectroscopy, on the other hand, is often known as Attenuated total reflection (ATR) and it is widely used in the IR region, (Fahrentfort, 1961) and (Harrick, 1967). In ATR, a crystal with high refractive index is placed in contact with the sample. The incidence angle for the illuminating beam is set greater than the critical value Θ_c (see Figure 6). Under these conditions all incident beams are completely reflected at the interface.

In ATR spectroscopy, the changes that take place in a totally reflected IR beam when it comes into contact with the sample are measured. In this technique, an IR beam is directed onto a crystal with a high refractive index. The sample is placed in close contact with the crystal. The refractive index for the crystal should be higher than the sample refractive index. The internal reflection creates an evanescent wave that interacts with the sample penetrating only 0.5-5 μm beyond the crystal. The wave is attenuated at wavenumbers where the sample absorbs. This attenuation as a function of the light energy is perceived by a detector and recorded as the ATR spectra.

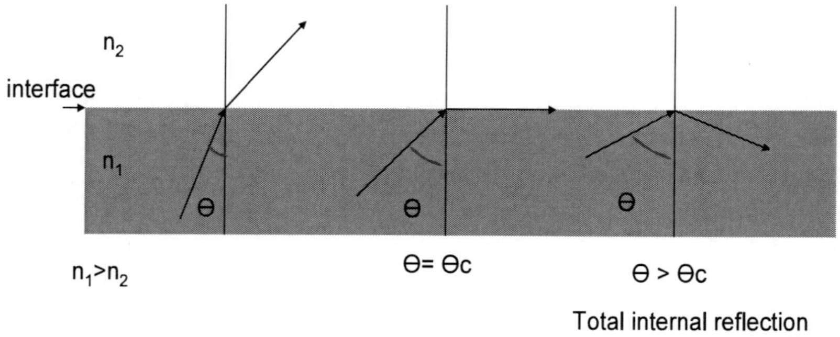

Figure 6. When a beam light strikes a medium boundary, part of the light is refracted and part is reflected. However, if the incident angle is greater than a critical angle Θ_c, the beam cannot cross the boundary and it is totally reflected. This total internal reflection is only possible if light travels from a medium with refractive index n_1 to a medium with refractive index n_2, being $n_1 > n_2$.

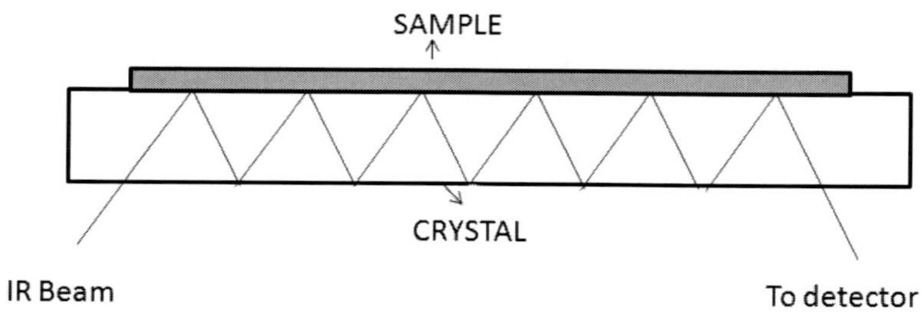

Figure 7. Schematic representation of an evanescent wave in an ATR set up.

Figure 7 represents the evanescent wave in a typical ATR experiment.

Different materials as germanium, diamond, zinc selenide (ZnSe), zinc sulfide (ZnS), Cd telluride (CdTe), etc., with refractive indexes varying usually from 2.2 to 4 (at 1000 cm^{-1}) are used as ATR crystals.

ATR is a non-destructive technique for different materials including liquids, solid materials, powders, surface layers, polymer films, gels, pastes. It presents an advantage for thick and dark colored materials which usually absorb too much radiation to be measured by IR transmission. (Maribeli, 1993), (Koulis et al., 2001)

In diffuse reflectance spectroscopy (external reflection), the electromagnetic radiation reflected by the surface is collected and analyzed. When this technique is applied in IR region with Fourier transform, it is named as diffuse reflectance infrared Fourier transform spectroscopy (DRIFTS).

5.3. Time-resolved Reflectance

In time-resolved spectroscopy, the sample is irradiated with a light source whose intensity fluctuates as a function of time. A pulsed LASER is generally used as the irradiation source. The LASER light is directed to the sample by means of a glass fiber. Diffusely reflected photons are then collected at a given distance from the source (around 40 mm) by a fiber optic bundle in contact with the sample. A two-stage microchannel plate photomultiplier tube may be used as detector and the pulse shape is recorded using time resolved photon counting methods. A small fraction of the incident light is coupled to the photomultiplier tube to account for on-line recording of the instrument response function.

One of the important characteristics of time-resolved diffuse reflectance spectroscopy is the capability to contain information dependent on depth in time-resolved curves. Photons collected at early times give information on the superficial layers while the photons collected at late times keep information on the deeper ones. (Steinbrink et al., 2001)

6. Applications of Reflectance Spectroscopy

6.1. Chromophore Concentrations from Reflected Light

Analysis of reflectance spectra has been used in bibliography to obtain the concentration of pigments and other absorbing substances non-destructively. During the last years, non invasive determination of the content of photosynthetic pigments has been particularly a focus of attention.

Near-infrared (NIR) reflectance spectroscopy has been used to predict the water content of oats (*Avena sativa*). It was also possible to estimate by this technique protein concentration in finaly-ground grains. (Rabault and Downey, 1990)

Gitelson and Merzlyak studied reflectance of higher plant leaves. They found that wavelengths with maximum sensitivity to chlorophyll concentration were found from 530 to 630 nm and near 700 nm. The ratios of reflectance in the NIR (above 750 nm) to that of 700 nm and to that of 550 nm (R_{NIR}/R_{700} and R_{NIR}/R_{550}) were directly proportional to chlorophyll concentration in leaves (Gitelson and Merzlyak, 1998).

Reflectance spectra from pink petals of *Rhododendron indicum* flowers showed absorption bands in the UV due to phenolic compounds, in the visible around 533 nm (due to anthocyanins) and in the near IR (1470, 1930 and 2500 nm) due to water. A linear correlation between the remission function at 533 nm and the anthocyanin concentration expressed as μmol per gram fresh weight has been published. (Iriel and Lagorio, 2009).

Non-destructive assessment of water and pigments in leaves was also performed using the remission function and Kubelka-Munk theory. Linear correlations have been found between the water content in mmol per cm^2 of leaves and the remission function at 1456 nm and between total chlorophyll in mmol per cm^2 and the remission function at 550 and 700 nm. (Cordon et al., 2010).

Reflectance and digital imaging have also been successfully used to develop a computational-vision methodology to infer the pigments concentration in plant leaves with spatial resolution (Cordon and Lagorio, 2010).

Lin et al. developed a reflectance-based model for estimating chlorophylls of fresh and water stressed leaves using the reflectance at the absorption bands of chlorophyll a and b and the red edge spectrum (Lin et al., 2013)

6.2. Remote Sensing, Plant Physiology and Ecological Studies

Sunlight reflected from different objects on the earth is used in the remote sensing of the environment. Different targets (water, grass, forests, bare soil, buildings) have different spectral reflectance signatures and can be differentiated from remote "observation". The reflected light from the earth is detected by optical remote sensing systems which can be basically one of these four types: panchromatic, multispectral, superspectral and hyperspectral. In panchromatic systems (IKONOS PAN, SPOT HRV-PAN9), the sensor works with a single detector sensitive to light within an extended wavelength range. In multispectral imaging systems (Landsat MSS, Landsat TM, SPOT HRV-XS, IKONOS MS) the sensor has a multichannel detector with a few spectral bands. Superspectral imaging systems have more channels (about ten) than a multispectral with narrower bandwidths. Finally, hyperspectral imaging (Hyper ion) acquires images in about one hundred spectral bands improving greatly wavelength resolution. (Avery and Berlin, 1992), (Leblon, 1997), (Lillesand and Keifer, 1994), (Mulder, 1987), (Sabins, 1994)

For plants, several reflectance indices have been developed for estimation of plant health. The normalized difference vegetation index (NDVI) for instance, is calculated as:

$$NDVI = \frac{R_{NIR} - R_{VIS}}{R_{NIR} + R_{VIS}} \quad (27)$$

where R_{VIS} is the reflectance value measured with a sensor channel sensitive to 580-680 nm and R_{NIR} the reflectance acquired by a sensor sensitive to wavelengths 725-1100 nm. Healthy plants usually display NDVI values around 0.7 while highly stressed plants or dry vegetation can show NDVI values as low as 0.1-0.2. NDVI values in the range 0-0.1 are not due to vegetation but to bare soils and rocks. Negative values can be obtained from the presence of clouds, rains or snow. (Kriegler et al., 1969), (Gates, 1980), (Crippen, 1990), (Sellers, 1985), (Myneni et al., 1995), (Holben, 1986), (Richardson and Wiegand, 1977), (Huete, 1988), (Kaufman and Tenre, 1992), (Pinty and Verstraete, 1992), (Deering, 1978)

The photochemical reflectance index (PRI) is another parameter very useful in monitoring plant primary productivity remotely and it is defined as a function of reflectance values at 571 and 531 nm (Gamon et al., 1992):

$$PRI = \frac{R_{570} - R_{531}}{R_{570} + R_{531}} \quad (28)$$

A relationship between PRI and light use efficiency by plants has been obtained. In fact, it is reported that this index assess the efficiency of a plant for using the absorbed photosynthetically active radiation in the photosynthetic process. (Peñuelas et al., 2011)

A water index WI, which allows determination of water content in plant leaves has also been defined in literature (Peñuelas et al., 1997):

$$WI = \frac{R_{900}}{R_{970}} \quad (29)$$

Several authors (Gao, 1996), (Zarco Tejada et al., 2002), (Link 3), suggested another index for water content determination from the canopy reflectance, the normalised difference water index, NDWI:

$$\mathrm{NDWI} = \frac{R_{860} - R_{1240}}{R_{860} + R_{1240}} \quad (30)$$

Remote sensing is also used for analysis of soils. Their reflectance is determined by the mineral composition, the moisture, the content in organic matter and the texture. Shepherd and Walsh have published an article on the development of Reflectance Spectral libraries for characterization of soils properties. (Shepherd and Walsh, 2002)

Dalal and Henry published the simultaneous determination of moisture, organic carbon, and total nitrogen by near infrared reflectance spectrophotometry of soils. (Dalal and Henry, 1986)

Sudduth and Hummel, determined organic matter, cation exchange capacity and moisture from NIR reflectance of soils. (Sudduth and Hummel, 1993).

In a recent article, a descriptive comparison of soil reflectance spectra and calibration models using multiple spectrometers was published. (Ge et al., 2011)

In ecological studies, the analysis of nutrient concentration in species is relevant. NIR reflectance spectroscopy is also a useful tool in this field. Chemical compounds containing C-H, N-H and O-H absorb in the region between 750 and 2500 nm. Crude protein in grasses usually correlates with absorbance at 2120-2160 nm (the spectral region 2100-2200 nm corresponds to N-H stretching in amide bonds). Phenolics compounds absorb around 1650 nm and distinction between rich and poor in phenolics plants may be obtained by comparison of reflectance spectra in this region. Cautions may be however be taken because the presence of some compounds may interfere in the determination. For instance, lignin absorbs around the same wavelength than phenolics and acts as interference. For this reason pre calibrations are necessary for each studied species to find the convenient correlation between the compound content and a spectral parameter. As a general procedure, a multivariate statistical model is developed to describe the relationship between the NIR spectra and the chemical components of the sample and then, the statistical model is used to predict the composition of unknown samples of similar types. An excellent review on this subject has been published by (Foley et al., 1998).

6.3. Diseases Diagnosis

Physicians have used visual examination from ancient times to infer the presence of anomalies in human's health. Quantitative and objective "observations" can be achieved by means of spectrosocpic techniques. Reflectance spectroscopy is one of these useful and non invasive methodologies. Reflectance spectroscopy may be applied whenever an anomaly induces a change in the tissue reflection properties. This change is often due to the formation of substances (absent in the health tissue) absorbing at different wavelengths.

It is also important that the excitation beam can reach the chromophore (placed in the tissue) to be excited. The penetration depth of light into a tissue, defined as the depth where the intensity of the excitation light is decreased to 1/e (e = 2.71828) of the incident intensity, depends on light wavelength. UV light is strongly absorbed by human tissues and as consequence; the penetration depth is usually very small (about few μm). On the other hand, NIR light is poorly absorbed and its penetration depth can be in the order of several mm. As a general rule and due to the short penetration depth of light, reflectance spectroscopy is used for diseases affecting structures near surfaces. (Randeberg, 2005)

Diffuse reflectance Spectroscopy has been used to determine the skin absorption and scattering coefficients from the visible to the near infra-red. The influence of collagen concentration on the optical properties of the skin has been studied and differences between normal skin and collagen defficient skin were detected on the base of differences in the Diffuse reflectance spectra. (Tseng et al., 2012, 2009a, 2009b and 2008), (Grant et al., 2009), (Tseng and Durkin, 2008).

Reflectance spectroscopy has also been used to monitor blood oxygenation. The test is based on the fact that hemoglobin without bound oxygen (deoxyhemoglobin) presents absorption maxima at 550 and 760 nm, displaying a dark red bluish color while Hemoglobin with bound oxygen (oxyhemoglobin) absorbs at 548 and 576 nm and is bright red. Reflectance spectra are so clearly different for the oxygenated and deoxygenated forms of hemoglobin allowing monitoring of their proportion in blood from the study of the skin spectra.

Methemoglobinemia, observed after overdoses of some drugs or poisoning, can also be detected by reflectance spectroscopy of skin. Methemoglobin is a brown oxidation product of hemoglobin, containing Fe(III) and absorbing at 404, 508 and 635 nm. (Randeberg, 2005)

Neonatal jaundice is also detectable by reflectance spectra. This anomaly is due to an increase in bilirrubin concentration form the breakdown of heme proteins (hemoglobin and myoglobin). These breakdown ussually takes place in the liver of newborns. Bilirrubin has an absorption peak at 460 nm displaying a yellow orange color. (Lodha et al., 2000), (Randeberg, 2005)

Reflectance spectroscopy and color coordinates have been used as an alternative to traditional histology for the diagnosis of liver pathologies from spectra of liver tissues. The accumulation of fat inside liver cells produces changes in their optical properties and can be detected by reflectance spectroscopy. (Randeberg, 2005)

6.4. Cultural Heritage Objects

Reflectance Spectroscopy can be used to detect dyes and fraud in artwork. In historical paintings, anthraquinone derivatives were frequently used as red dyes. These dyes were extracted from insects as cochineals (carminic acid) or kermes (kermesic acid) and from plants (madder dyes as alizarin or purpurin) (Schweppe and Winter, 1997), (Schweppe and Roosen-Runge, 1986).

Using Fibre Optic Reflectance Spectroscopy (FORS) in the region between 200 and 1000 nm, it was possible to detect the presence of anthraquinones based dyes in paintings and in some cases to identificate the dye origin (insect or plant). (Leona and Winter, 2001), (Bacci et al., 2000), (Bisulca et al., 2008)

In the Fibre Optic Reflectance Spectroscopy optical fibers are used both to guide the incoherent radiation emitted from a light source to the sample, and to collect the reflected light and send it to a spectroanalyzer. The UV-VIS-NIR spectrum was reported to be particularly useful for determining the chemical composition of the dyes used in the painting, and for monitoring superficial alterations to the artwork's color. (Cucci et al., 2011), (Bacci et al., 2008), (Dupont at al., 2008), (Bacci et al., 1991), (Leona and Winter, 2001), (Dupuis et al., 2002), (Bacci et al., 2003), (Cucci et al., 2013), (Cucci et al., 2011), (Casini et al., 2005)

6.5. Other Applications

In addition to the applications described, numerous other uses may be found in literature for reflectance measurements: identification of components in toilet articles and household products by attenuated total reflectance (Puttnam et al., 1965); post harvesting monitoring of fruits during storage by time-resolved reflectance techniques (Vanoli et al., 2011), study of surface coatings (surface treated metals, paints, polymers) using specular reflectance measurements (Unemura, 2002), (Kim, 2007), (Griffiths and de Haseth, 1998), (Link 4) among many others.

CONCLUSION

Reflectance spectroscopy (RS) is an analytical technique that offers many advantages: minimal or no sample preparation is needed in most cases, samples may be analysed in a non-destructive way with high sensitivity, RS is applicable to remote sensing, it is suitable for scattering samples and for the analysis of coatings. Different fields of science have benefited from the various alternatives of reflectance techniques: industrial production of paints, coatings and textiles, biomedicine, agriculture, satellite monitoring, etc. RS is an extremely valuable tool nowadays and it is expected to increase its importance in the coming years.

REFERENCES

Allen, W. A., Richardson, A., 1968. Interaction of light with a plant canopy, *J. Opt. Soc. Am.* 58, 1023–1028.

Avery, T. and Berlin, G., 1992. *Fundamentals of Remote Sensing and Airphoto Interpretation.* 5th edition. Maxwell MacMillan Company. Toronto.

Bacci, M., Orlando, A., Picollo, M., Radicati, B., Lanterna, G., 2000. Colour analysis of historical, red lakes using non-destructive reflectance spectroscopy. Compatible Materials for the Protection of Cultural Heritage, in *PACT* 58, 21-35.

Bacci, M., Baldini, F., Carlà, R., Linari, R., 1991. A color analysis of the Brancacci Chapel frescoes, *Appl. Spectrosc.* 45, 26-31.

Bacci, M., Cucci, C., Mencaglia, A. A., Mignani, A. G., 2008. Innovative sensors for environmental monitoring in museums, *Sensors* 8, 1984-2005.

Bacci, M., Casini, A., Cucci, C., Picollo, M., Radicati, B., Vervat, M., 2003. Non-invasive spectroscopic measurements on the Il ritratto della figliastra by Giovanni Fattori: identification of pigments and colourimetric analysis, *J. Cult. Heritage* 4, 329-336.

Bisulca, C., Picollo, M., Bacci, M., Kunzelman, D., 2008. UV-VIS-NIR Reflectance Spectroscopy of red lakes in paintings, *9th International Conference on NDT of Art*, Jerusalem Israel, 25-30 May 2008, 1-8.

Cabrera, M. I., Alfano, O. M., Cassano, A. E., 1996. Absorption and scattering coefficients of titanium dioxide particle suspensions in water. *J. Phys. Chem.* 100, 20043-20050.

Casini, A., Bacci, M., Cucci, C., Lotti, F., Porcinai, S., Picollo, M., Radicati, B., Poggesi, M., Stefani, L., 2005. Fiber optic reflectance spectroscopy and hyper-spectral image spectroscopy: two integrated techniques for the study of the Madonna dei Fusi, *Proc. SPIE 5857*, p. 58570M, doi:10.1117/12.611500.

Cordon, G. B. and Lagorio, M. G., 2007. Absorption and scattering coefficients. A biophysical-chemistry experiment using reflectance spectroscopy, *J. Chem. Educ.*, 84, 1167-1170.

Cordon, G. B., Gismondi, S., Nievas, A. V., Lagorio, M. G., 2010. Non-destructive assessment of water and pigments in leaves from the remission function using the Kubelka-Munk theory, *Proceedings of the AIC Color and Food, Interim Meeting of the International Color Association*, 397-400.

Cordon, G. B. and Lagorio, M. G., 2010. Sensing chlorophyll and anthocyanin concentration in leaves with spatial resolution from digital image. *Proceedings of the AIC Color and Food, Interim Meeting of the International Color Association,* 425-428.

Crippen, R. E., 1990. Calculating the vegetation index faster, *Remote Sensing of Environment,* 34, 71-73.

Cucci, C., Bacci, M., Picollo, M., Olmi, R., 2011. Non invasive techniques for identification and characterisation of polymers in contemporary artworks, *Future Talks 009 The Conservation of Modern Materials in Applied Arts and Design*, p. 178-184, Die Neue Sammlung, Munich.

Cucci, C., Bigazzi, L., Picollo, M., 2013. Fibre optic reflectance spectroscopy as a non-invasive tool for investigating plastics degradation in contemporary art collections: a methodological study on an expanded polystyrene artwork, *J. Cult. Heritage,* 14, 290-296.

Cucci, C., Casini, A., Picollo, M., Poggesi, M., Stefani, L., 2011. Open issues in hyperspectral imaging for diagnostics on paintings: when high spectral and spatial resolution turns into data redudancy, *Proc. SPIE 8084*, p. 808408, doi:10.1117/12.889460.

Dalal, R. C., Henry, R. J., 1986. Simultaneous determination of moisture, organic carbon, and total nitrogen by near infrared reflectance spectrophotometry. *Soil Sci. Soc. Am. J.,* 50, 120-123.

Deering, D.W., 1978. Rangeland reflectance characteristics measured by aircraft and spacecraft sensors. Ph.D. Diss. Texas A&M Univ., College Station, 338p.

Duncan, D. R., 1940. The colour of pigment mixtures, *Proc. Phys. Soc.* 52, 390-401.

Duncan, D. R., 1949. The colour of pigment mixtures, *J. Oil Colour Chem. Assoc.* 32, 296-391.

Dupont, A.-L., Cucci, C., Loisel, C., Bacci, M., Lavedrine, B., 2008. Development of LightCheck Ultra, a novel dosimeter for monitoring lighting conditions of highly photosensitive artefacts in museums, *Stud. Conserv.* 53, 49-72.

Dupuis, G., Elias, M., Simonot, L., 2002. Pigment identification by fiber-optics diffuse reflectance spectroscopy, *Appl. Spectrosc.* 56, 1329-1336.

Fahrenfort, J., 1961. Attenuated total reflection: A new principle for the production of useful infra-red reflection spectra of organic compounds, *Spectrochimica Acta*, 17, 698-709.

Foley, W. J., McIlwee, A., Lawler, I., Aragones, L., Woolnough, A. P., Berding, N., 1998. Ecological applications of near infrared reflectance spectroscopy -a tool for rapid, cost-effective prediction of the composition of plant and animal tissues and aspects of animal performance Oecologia, 116, 293-305.

Gamon, J., Peñuelas, J. and Field, C., 1992. A narrow-waveband spectral index that tracks diurnal changes in photosynthetic efficiency. *Remote Sensing of Environment*, 41, 35-44.

Gao, B. C., 1996. NDWI-a normalized difference water index for remote sensing of vegetation liquid water from space. *Remote Sensing of Environment,* 58, 257- 266.

Gates, D. M., 1980. Biophysical Ecology, Springer-Verlag, New York, 611 p.

Ge, Y, Morgan, C. L. S., Grunwald, S., Brown, D. J., Sarkhot, D. V., 2011. Comparison of soil reflectance spectra and calibration models obtained using multiple spectrometers, *Geoderma* 161, 202-211.

Gitelson, A. A. and Merzlyak, M. N., 1998. Remote sensing of chlorophyll concentration in higher plant leaves, *Adv. Space Res.*, 22, 5, 689-692.

Gu X, Zhang Q, Bartlett M, Schutz L, Fajardo LL, Jiang H., 2004. Differentiation of cysts from solid tumors in the breast with diffuse optical tomography, *Acad. Radiol.*, 14, 53-60.

Griffiths, P. R., de Haseth, J. A., 1998. *Fourier Transform Infrared Spectrometry,* New York, John Wiley & Sons.

Grant, A. M., Sry, K., Saager, R., Ayers, F., Pfefer, T. J., Kelly, K. M., Tseng, S-H, Durkin, A. J., 2009. Diffuse optical spectroscopy of melanoma-simulating silicone phantoms, *Proc. of SPIE* Vol. 7187, 718702.

Harrick, N. J., 1967. Internal Reflection Spectroscopy. New York, John Wiley & Sons, Int.

Holben, B. N., 1986. Characteristics of Maximum-Value Composite Images from Temporal AVHRR Data, *International Journal of Remote Sensing*, 7, 1417-1434.

Huete, A. R., 1988. A soil-adjusted vegetation index (SAVI), *Remote Sensing of Environment*, 25, 53-70.

Iriel, A. and Lagorio, M. G., 2009. Biospectroscopy of Rhododendron indicum flowers. Non-destructive assessment of anthocyanins in petals using a reflectance-based method, *Photochem. Photobiol. Sci*, 8, 337-344.

Kaufman, Y. J. and Tanre, D., 1992. Atmospherically resistant vegetation index (ARVI) for EOS-MODIS, *Proc. IEEE Int. Geosci. and Remote Sensing Symp.92*, IEEE, New York, 261-270.

Khoshhesab, Z. M., 2012. Reflectance IR Spectroscopy, Infrared Spectroscopy - Materials Science, Engineering and Technology, Prof. Theophanides Theophile (Ed.), ISBN: 978-953-51-0537-4, InTech, Available from: http://www.intechopen.com/books/infrared-spectroscopy-materials-science-engineering-and-technology/fundamental-of-reflectance-ir-spectroscopy.

Kim, J., 2007. Investigations of Thiolated Self-Assembled Monolayers on Gold Substrates by FTIR with Specular Reflectance, *PIKE Technologies Application Note*, www.piketech.com.

Kortüm, G., 1969. Reflectance spectroscopy: principles, methods, applications, New York, Springer-Verlag, United States.

Koulis, C. V., Reffner, J. A., John, B. A., Bibby, A. M., 2001. Comparison of transmission and internal reflection infrared spectra of cocaine. *J. Forensic Science*, 46, 822.

Kriegler, F. J., Malila, W. A., Nalepka, R. F., and Richardson, W., 1969. Preprocessing transformations and their effects on multispectral recognition, *Proceedings of the Sixth International Symposium on Remote Sensing of Environment*, 97-131.

Lagorio, M. G., 2004. Why do Marbles become paler on grinding? Reflectance Spectroscopy, Color and particle size, *J. Chem. Educ.*, 81, 1607-1611.

Lambert, J. H., 1760. Photometria, Augsburg, German edition by Anding Leipzig 1892.

Leblon, B., 1997. The Remote Sensing Core Curriculum; "Optical Properties of Vegetation and Soil". Remote Sensing and GIS Laboratory, Faculty of Forestry and Environmental Management. University of New Brunswick, Frederick (NB), Canada. E3B 6C2. Linked through the University of Texas Website: http://www.utexas.edu/depts/grg/virtdept/contents.html.

Leona, M. and Winter, J., 2001. Fiber Optics Reflectance Spectroscopy: A Unique tool for the investigation of paintings, *Studies in Conservation* 46, 153-162.

Lillesand, T. M. and Keifer, R.W., 1994. *Remote Sensing and Image Interpretation,* Third Edition. John Wiley and Sons, Inc. Toronto.

Lin, C., Popescu, S. C., Huang, S. C., Chen, Y. C., Chang, P. T., and Wen, H. L., 2013. A novel reflectance-based model for evaluating chlorophyll concentration of fresh and water-stressed leaves, *Biogeosciences Discuss.*, 10, 17893-17937.

Lodha, R., Deorari, A. K., Jatana, V., Paul, V. K., 2000. Non-Invasive Estimation of Total Serum Bilirubin by Multi-Wavelength Spectral Reflectance in Neonates, *Indian Pediatrics*, 37, 771-775.

Maribeli, F. M., 1993. Principles, Theory, and Practice of Internal Reflection Spectroscopy,

Internal Reflection Spectroscopy-Theory and Applications, ed. F. M. Mirabella, Jr., Marcel Dekker, Inc. New York.

Myneni, R. B., Hall, F. G., Sellers, P.J. and Marshak, A.L., 1995. The interpretation of spectral vegetation indexes, *IEEE Transactions on Geoscience and Remote Sensing*, 33, 481-486.

Mulder, M. A., 1987. Application of Remote Sensing to Soils, *Developments in Soil Science* 15. Elsevier, New York.

Peñuelas, J., Garbulsky, M. F. and Filella, I., 2011. Photochemical reflectance index (PRI) and remote sensing of plant CO2 uptake, *New Phytologist*, 191, 596-599.

Peñuelas, J., Piña, J., Ogaya, R. and Flella, I., 1997. Estimation of plant water concentration by the reflectance Water Index WI (R900/R970), *Int. J. Remote Sensing*, 18, 13, 2869- 2875.

Pinty, B. and Verstraete, M. M., 1992. GEMI: A non-linear index to monitor global vegetation from satellites, *Vegetation*, 101, 15-20.

Puttnam, N. A., Lee, S. and Baxter, B. H., 1965, Application of Attenuated Total Reflectance IR spectroscopy to toilet articles and household products, Qualitative analysis, *J. Soc.Cosmetic Chemists*, 1, 607-615.

Rabault, J. L. and Downey, G., 1990. Rapid Analysis of Oats for Moisture and Proteins by Near-Infrared Reflectance Spectroscopy, *Irish Journal of Food Science and Technology,* 14, 85-93.

Randeberg, L. L., 2005. Diagnostic applications of diffuse reflectance spectroscopy, DOCTORAL THESIS submitted to the Department of Electronics and Telecommunications, Norwegian University of Science and Technology, in partial fulfillment of the requirements for the degree of doktor ingeniør Trondheim, Doctoral Theses at NTNU, 2005:100 ISBN 82-471-7077-9.

Richardson, A. J. and Wiegand, C. L., 1977. Distinguishing vegetation from soil background information, *Photogrammetric Engineering and Remote Sensing*, 43, 1541-1552.

Sabins, F. F., 1994. *Remote Sensing: Principles and Interpretation.* John Wiley and Sons. Toronto.

Schweppe, H. and Winter, J., 1997. *Madder and alizarin, Artists' Pigments: A Handbook of Their History and Characteristics* Vol. 3, ed. R. L. Feller National Gallery of Art, Washington DC, 109-142.

Schweppe, H., and Roosen-Runge, H., 1986, Carmine-Cochineal Carmine and Kermese Camine, in Artists' Pigments: *A Handbook of Their History and Characteristics,* Vol. 1, ed. R. L. Feller, National Gallery of Art, Washington DC, 255-283.

Sellers, P. J., 1985. Canopy reflectance, photosynthesis, and transpiration, *International Journal of Remote Sensing*, 6, 1335-1372.

Shepherd, K. D. and Walsh, M. G., 2002. Development of Reflectance Spectral Libraries for Characterization of Soil Properties, *Soil Sci. Soc. Am. J.*, 66, 988-998.

Steinbrink, J., Wabnitz, H., Obrig, H., Villringer, A., and Rinneberg, H., 2001. Determining changes in NIR absorption using a layered model of the human head, *Phys. Med. Biol.* 46, 879-896.

Stokes, G. G., 1860. On the intensity of the light reflected from or transmitted through a Pile of Plates, *Proceedings of the Royal Society of London*, 11, 545-557.

Sudduth, K. A., Hummel, J. W., 1993. Soil organic matter, CEC, and moisture sensing with a portable NIR spectrophotometer. *Trans. ASAE*, 36, 1571-1582.

Tseng, S-H., Hsu, C-K, Lee, J. Y-Y., Tzeng, S-Y, Chen, W-R. and Liaw, Y-K., 2012. Noninvasive evaluation of collagen and hemoglobin contents and scattering property of in vivo keloid scars and normal skin using diffuse reflectance spectroscopy: pilot study', *J. Biomed. Opt.*,17, 077005.

Tseng, S-H, Bargo, P., Durkin, A. and Kollias, N., 2009a. Chromophore Concentrations, Absorption and Scattering Properties of Human Skin In-vivo, *Optics Express*, 17, 14599-14617.

Tseng, S-H, Hayakaw, C., Spanier, J. and Durkin, A. J., 2009b. Investigation of a probe design for facilitating the uses of the standard photon diffusion equation at short source-detector separations: Monte Carlo simulations, *Journal of Biomedical Optics*, 14, 054043.

Tseng, S-H and Durkin, A. J., 2008. In-vivo Determination of Skin Near-Infrared Optical Properties Using Diffuse Optical Spectroscopy, *Journal of Biomedical Optics*, 13, 014016.

Tseng, S-H, Hayakawa, C., Spanier, J. and Durkin, A. J., 2008. Determination of Optical Properties of Superficial Volumes of Layered Tissue Phantoms", *IEEE Transactions on Biomedical Engineering*, 55, 335-339.

Unemura, J., 2002. Reflection-Absorption Spectroscopy of Thin Films on Metallic Substrates, Handbook of Vibrational Spectroscopy, Vol. 2, 982-998, Chalmers, J. M. , Griffiths, P. R. Eds., New York, John Wiley & Sons Ltd.

Vanoli, M., Rizzolo, A., Grassi, M., Farina, A., Pifferi, A., Spinelli, L., Torricelli, A., 2011. Time-resolved reflectance spectroscopy nondestructively reveals structural changes in 'Pink Lady®' apples during storage, *Procedia Food Science*, 1, 81-89.

Wendlandt, W. W. M. and Hecht, H. G., 1966. In *Reflectance Spectroscopy;* Intersciences Publishers: New York.

Zarco-Tejada, P. J., Rueda, C.A., Ustin, S.L., 2003. Water content estimation in vegetation with MODIS reflectance data and model inversion methods, *Remote Sensing of Environment*, 85, 109-124.

Links

http://www.ssi.shimadzu.com/products/product.cfm?product=uv_accy_solid_guide1, accessed in 2014.

http://www.perkinelmer.com/Catalog/Category/ID/Integrating%20Spheres, accessed in 2014.

http://ecologia.ib.usp.br/lepac/bie5759/NDWI.pdf, accessed in 2014.

http://www.piketech.com/skin/fashion_mosaic_blue/application-pdfs/AnalysisCoatingsMetallicSubstratesSpecularReflectanceFTIRSpectroscopy.pdf, accessed in 2014.

In: Optical Spectroscopy
Editor: Nicolae Tomozeiu

ISBN: 978-1-63321-197-1

Chapter 3

PHOTOREFLECTANCE SPECTROSCOPY OF FRANZ-KELDYSH OSCILLATIONS FROM SEMICONDUCTOR HETEROSTRUCTURES FOR ELECTRONIC AND OPTOELECTRONIC DEVICES AND COMPONENTS

Hideo Takeuchi*

Associate Professor, Ph.D., Department of Applied Physics,
Graduate School of Engineering, Osaka City University
Sumiyoshi, Osaka, Japan

ABSTRACT

The present chapter demonstrates that photoreflectance spectroscopy, with high accuracy, has effectiveness to diagnose and develop semiconductor epitaxial structures for electronic, optical, and optoelectronics devices. We focus our attention on Franz-Keldysh oscillations appearing in the photoreflectance spectra, which provide information on the built-in electric field in the epitaxial structures. We present a theory that takes account of the interference effect of the probe-light beam, which is hardly taken into account. Using the above-

* E-mail address: Takeuchi@a-phys.eng.osaka-cu.ac.jp; Tel: +81-6-6605-2174.

mentioned theory, we discuss the line shape of the Franz-Keldysh oscillations. We show the appropriateness that is essential to precisely evaluate built-in electric field. We present various case studies using the photoreflectance measurement of the Franz-Keldysh oscillations in the heterostructures.

I. INTRODUCTION

Photoreflectance spectroscopy and/or electroreflectance spectroscopy have the ability of characterizing optical properties of semiconductors. Optical transition energies have been intensively investigated using the electroreflectance measurement because dielectric functions exhibit large changes at the critical points of the wave vector space. It is noteworthy that the obtained optical transition parameters are generally used for the reference value. The photoreflectance and electroreflectance measurements are categorized into the modulation spectroscopic measurement [1]. In performing the modulation spectroscopic measurements, samples are subjected to relatively small periodical external fields. At the same time, the probe light beam is illuminated to the samples. The periodical external field modulates dielectric functions; therefore, the reflected probe beam contains the dielectric function components modulated by the periodical external fields. The modulated reflection-beam component is usually detected with use of lock-in detection. The lock-in detection is highly sensitive to the periodical signals because of its large dynamic range and signal-to-noise ratio. Since, as mentioned above, the dielectric functions drastically changes at the critical points of the energy band structure, the modulated reflection-beam component provide a large amount of information on the optical transitions at the critical points. This is a fundamental advantage of the modulation spectroscopic measurement.

In the electroreflectance measurement, the modulation is driven by an electric field. The electrodes are placed on the samples. The small periodical (alternative) electric field is applied to the sample through the electrodes and modulates the dielectric function of the sample. Comparing with the magnetic fields, the electric field is easily available. This is an advantage of the electroreflectance measurement.

The photoreflectance measurement is derived from the electroreflectance measurement. Figure 1 schematically shows the photoreflectance mechanism semiconductors, the built-in electric field is formed by the phenomenon of surface Fermi level pinning. The built-in electric field is also produced by the *p-n* junction. The photoreflectance mechanism is based on a pump light beam, of a photon energy which is larger than fundamental transition energy. This pump beam generates carriers. The photogenerated carriers modulate the built-in electric field through the Coulomb screening effects. In the photoreflectance measurement, the modulation is caused by only the pump-beam illumination; namely, the photoreflectance measurement has an aspect of being contactless and hence nondestructive. The photoreflectance measurement is, therefore, applicable to the non-destructive evaluation of the semiconductors. This is the reason why the photoreflectance measurement has been attracting much attention in the characterization of epitaxial structures for various semiconductor components and devices. In the case where the built-in electric field strength is higher than about 5.0~10 kV/cm, the photoreflectance spectra exhibit the oscillation patterns, the Franz-Keldysh oscillations. Consequently, the appearance of the Franz-Keldysh oscillations readily connects with the presence of the built-in electric field. The built-in electric field is governed by the epitaxial structure. Accordingly, the analysis of the Franz-Keldysh oscillations is useful to clarify the potential structure of the epitaxial structure. For example, Pollak et al., applied the measurements of Franz-Keldysh oscillations to the characterization of the epitaxial structures for the heterojunction bipolar transistors [2,3].

In the present chapter, we demonstrate that the photoreflectance measurement of the Franz-Keldysh oscillations is effective to precisely evaluate the semiconductor epitaxial structures. In Section II, we introduce the experiment setup of the photoreflectance measurement. In Section III, as an introduction on the Franz-Keldysh oscillations, we describe the theory on the Franz-Keldysh oscillations in the bulk single crystals and simple epitaxial structures, in which the interference effects of the probe beam are absent. Here, we show a photoreflectance spectrum of the undoped GaAs/n-type GaAs (*i*-GaAs/*n*-GaAs) structure. In Section IV, we treat the Franz-Keldysh oscillations that originate from the epitaxial structures influenced by the presence of the probe-beam interference effects. We compare the experimental line shape with calculated line shape of the Franz-Keldysh oscillations from the *i*-GaAs/*n*-GaAs with the SiN_x cover film. We also demonstrate the experimental and calculated line shape of the Franz-Keldysh oscillations from two $Al_xGa_{1-x}As$/GaAs heterojunction bipolar transistor epitaxial structures In Section V, we demonstrate the high sensitivity of the Franz-Keldysh oscillations to the surface morphology of the $Al_xGa_{1-x}N$/GaN heterostructures for high electron mobility transistors. In Section VI, we demonstrate the simultaneous estimation of the electric field strength and the direction of the electric field along the growth direction, using the photoreflectance and terahertz spectroscopic measurements. In Section VII, we also demonstrate that the measurement of the Franz-Keldysh oscillations is applicable to the design and optimization of the terahertz emitters. Finally, we summarize the present chapter.

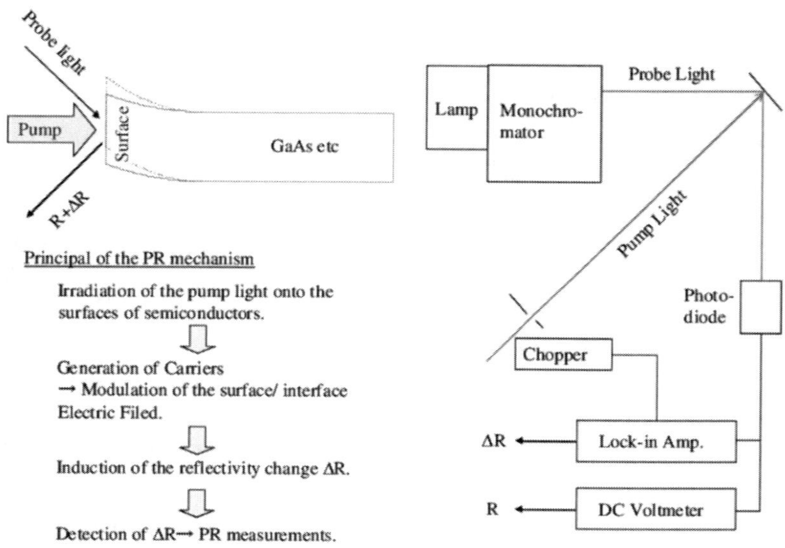

Figure 1. Schematic concept of the photoreflectance spectroscopic measurement. In this figure, the abbreviation of PR means photoreflectance.

II. Experimental Setup for the Photoreflectance Measurements

Figures 2 and 3 show the experiment setup for the optical alignment and electronic instruments used for the photoreflectance measurement. The source of the probe beam is usually the tungsten halogen lamp light dispersed by the single monochromator. In measuring wide band-gap semiconductors, a xenon lamp is sometimes used. The optical chopper periodically modulates the pump beam from the continuous wave laser. The chopping frequency ranges from 10^2 to 10^3 Hz. Instead of the continuous wave laser, the pulsed laser light is

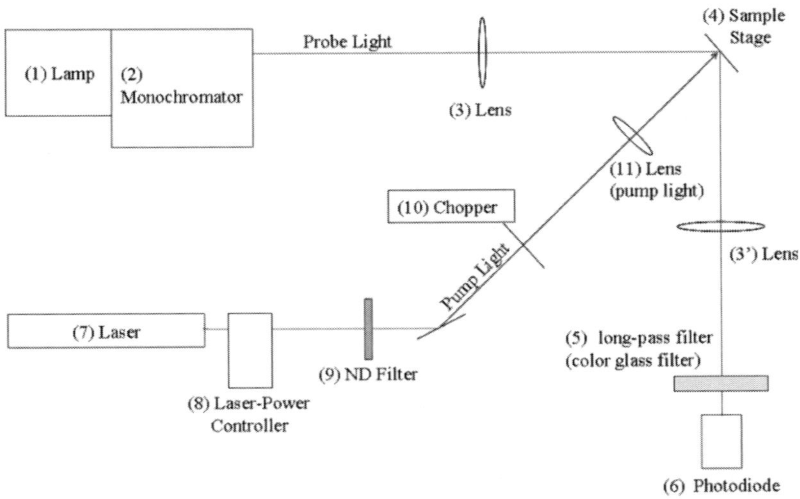

Figure 2. The optical alignment for photoreflectance spectroscopy.

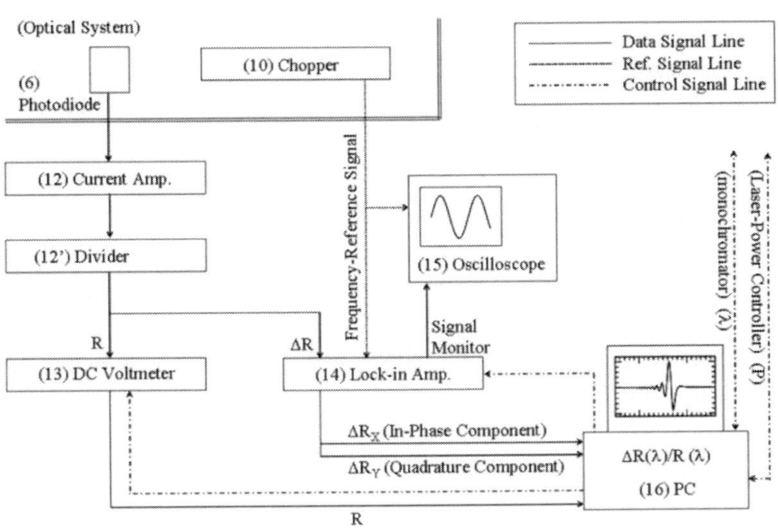

Figure 3. Diagram of the signal processes and the instrument control system for the photoreflectance measurement. Single lock-in amplifiers are also applicable to the photoreflectance measurement, instead of the two-phase lock-in amplifier.

applicable. The reflected probe beam is collected into the photodiode. The photocurrent generated in the detector is converted to the voltage with use of the current amplifier. The converted signal passes the divider: low pass filter, which extracts the modulated component from the total signal. The modulated component is recorded with use of the lock-in detection. As shown in Figures 2 and 3, the measurement system is quite simple. In fact, single monochromators are applicable to the photoreflectance measurement. The measurement of the Franz-Keldysh oscillations can be performed at room temperature, so that the photoreflectance measurement is relatively convenient.

III. Franz-Keldysh Oscillations Appearing in Bulk Single Crystals and Simple Epitaxial Structures: In the Absence of the Effects of the Interference Effects of the Probe Light Beam

In advance to discuss the Franz-Keldysh oscillations in epitaxial structures, we describe the theory on the Franz-Keldysh oscillations. The theories on the Franz-Keldysh oscillations were established by Franz [4], Keldysh, [5, 6], Aspnes [7-9], Pollak [10], and Cardona [1]. Here, we describe the characteristics of the Franz-Keldysh oscillations in bulk single crystal with a uniform and derive a simple asymptotic line-shape function.

The presence of the uniform built-in electric field lowers the crystal symmetry along the direction of the built-in electric field. It is fundamental to trigger the appearance of the Franz-Keldysh oscillations. Taking account of the built-in electric field, the Hamiltonian of the material system is written with use of Equation (1) [11]:

$$\left(-\frac{\hbar^2}{2\mu}\nabla_r^2 - eFz\right)\phi(\vec{r}) = E_r\phi(\vec{r}) \tag{1}$$

where the quantities μ, F, and E_r are the reduced mass at the critical point, electric field strength and eigen energy, respectively. The z axis is the direction of the electric field, and ϕ(r) is the envelope of the wave function. The solution of the envelope function in Equation (1) is, therefore, expressed by the following equation:

$$\phi(\vec{r}) = u(x, y)\varphi(z) \tag{2a}$$

The function $u(x,y)$

$$u(x, y) = \frac{1}{\sqrt{N}} \exp\left[-i\left(k_x x + k_y y\right)\right] \tag{2b}$$

The eigen energy E_r is given by

$$E_r = E_0 + \frac{\hbar^2\left(k_x^2 + k_y^2\right)}{2\mu} + E_z \tag{3}$$

Here, the fundamental transition energy E_0 is included in Equation (3). Equation (1) is modified to be simple with changing variables.

$$-\frac{\eta^2}{2\eta^2}\varphi(\eta) = \eta\varphi(\eta) \tag{4}$$

The change of the variable is performed with use of Equation (4):

$$\eta = \frac{E_z}{\hbar\Theta} - z\left(2\frac{\mu e}{\hbar^2}F\right)^{1/3} \tag{5}$$

Here, Equation (5) contains the important parameter characterizing the Franz-Keldysh oscillations: the electro-optic energy $\hbar\Theta$:

$$\hbar\Theta = \left(\frac{e^2F^2\hbar^2}{2}\right)^{1/3} \tag{6}$$

The solution of Equation (4) is obtained taking account of the boundary condition:

$$\lim_{z\to\pm\infty} \varphi(z) = 0 \tag{7}$$

The solution is

$$\varphi(\eta) = \frac{(eF)^{1/2}}{\hbar\Theta}\mathrm{Ai}(\eta) \tag{8}$$

The function Ai(η) is the Airy A function. In order to calculate of the imaginary part of $\varepsilon_i(\omega, F)$, here, we limit to treat in the case of the M_0 critical point [11]. The dielectric function of $\varepsilon_i(\omega, F)$ is obtained as the following equation:

$$\varepsilon_i(\omega, F) = \frac{2\pi e^2 (2\mu)^3 |P|^2 (\hbar\Theta)^{1/2}}{m^2\hbar(\hbar\omega)^2}\mathrm{Ai}(\eta)\int_{\eta_0}^{-\infty}\mathrm{Ai}^2(\eta)d\eta \tag{9a}$$

where

$$\eta_0 = \frac{E_0 - \hbar\omega}{(\hbar\Theta)^2} \tag{9b}$$

In Equation (9a), the quantity $|P|^2$ is the oscillator strength. The following relation exists in the Airy A function:

$$\int_{\eta_0}^{-\infty}\mathrm{Ai}^2(\eta)d\eta = \eta\mathrm{Ai}^2(\eta) - (\mathrm{Ai}'(\eta))^2 \tag{10}$$

Thus, the modulated component of the imaginary part of the dielectric function is expressed as

$$\begin{aligned}\Delta\varepsilon_i(\omega, F) &= \varepsilon_i(\omega, F) - \varepsilon_i(\omega, 0) \\ &= \frac{2e^2(2\mu)^{3/2}|p|^2(\hbar\Theta)^{1/2}}{m^2\hbar(\hbar\omega)^2} f\left(\frac{E_0 - \hbar\omega}{\hbar\Theta}\right)\end{aligned} \tag{11}$$

where

$$f(\varsigma) \equiv \pi\left[\mathrm{Ai}'(\varsigma)^2 - \varsigma\mathrm{Ai}^2(\varsigma)\right] - (\varsigma)^{1/2}U(-\varsigma) \tag{12}$$

The function U(ζ) is the Heaviside function:

$$U(\varsigma)=1 \ for \ \varsigma \geq 0 \\ \quad =0 \ for \ \varsigma < 0 \tag{13}$$

The real part of the modulated dielectric function is obtained using the Kramers-Kronig transform:

$$\Delta\varepsilon_r(\omega, F)=\frac{2e^2(2\mu)^{3/2}|p|^2(\hbar\Theta)^{1/2}}{m^2\hbar(\hbar\omega)^2}g\left(\frac{E_0-\hbar\omega}{\hbar\Theta}\right) \tag{14}$$

Here the function $g(\zeta)$ is

$$g(\varsigma)\equiv\pi[\mathrm{Ai'}(\varsigma)\mathrm{Bi'}(\varsigma)-\varsigma\mathrm{Ai}(\varsigma)\mathrm{Bi}(\varsigma)]+(\varsigma)^{1/2}U(-\varsigma) \tag{15}$$

The function $\mathrm{Bi}(\zeta)$ is the Airy B function. Since we have obtained the dielectric functions in the presence of the electric field F, we can calculate the photoreflectance spectrum $\Delta R/R$, which is obtained with use of the pump beam for the modulation. The photoreflectance spectrum is given by

$$\frac{\Delta R}{R}=\beta_r\Delta\varepsilon_r+\beta_i\Delta\varepsilon_i \tag{16}$$

The quantities β_r and β_i are the Seraphin coefficients [12]. In usual, at the fundamental transition range, Equation (16) is approximated and, as a result, is expressed by

$$\frac{\Delta R}{R}\approx\beta_r\Delta\varepsilon_r \tag{17}$$

In order to simply analyze the Franz-Keldysh oscillations in the presence of the uniform built-in electric field, it is necessary to modify Equation (17) with use of the asymptotic functions of $g(\zeta)$ and $f(\zeta)$. The asymptotic functions of $g(\zeta)$ and $f(\zeta)$ are expressed by $g_{asympt}(\zeta)$ and $f_{asympt}(\zeta)$ as follows:

$$f_{asympt}(\varsigma)=\frac{1}{32}\varsigma^{-5/2}-\frac{1}{4}\varsigma^{-1}\cos\left(\frac{4}{3}\varsigma^{3/2}\right) \\ \approx-\frac{1}{4}\varsigma^{-1}\cos\left(\frac{4}{3}\varsigma^{3/2}\right) \tag{18}$$

and

$$g_{asympt}(\varsigma) = \frac{1}{4}\varsigma^{-1}\sin\left(\frac{4}{3}\varsigma^{3/2}\right) \tag{19}$$

The asymptotic form of Equation (17), $(\Delta R/R)_{\text{asympt}}$ is, therefore, given by

$$\begin{aligned}\left(\frac{\Delta R}{R}\right)_{asympt} &\approx \beta_r \Delta\varepsilon_r \\ &\propto \frac{1}{4(\hbar\omega)^2}\left(\frac{\hbar\omega - E_0}{\hbar\omega}\right)\sin\left(\frac{4}{3}\left(\frac{\hbar\omega - E_0}{\hbar\Theta}\right)^{3/2}\right) \\ &= \frac{1}{4(\hbar\omega)^2}\left(\frac{\hbar\omega - E_0}{\hbar\omega}\right)\cos\left(\frac{4}{3}\left(\frac{\hbar\omega - E_0}{\hbar\Theta}\right)^{3/2} + \frac{\pi}{2}\right)\end{aligned} \tag{20}$$

The *j*-th extremum photon energy of the Franz-Keldysh oscillation, the peaks and/or dips $\hbar\omega$ in the photoreflectance spectrum is fulfilled with the following condition, where *j* is integers:

$$\frac{4}{3}\left(\frac{\hbar\omega - E_0}{\hbar\Theta}\right)^{3/2} + \frac{\pi}{2} = j\pi \tag{21}$$

Equation (21) can be transformed to be:

$$\begin{aligned}\hbar\omega &= \hbar\Theta\left[\frac{3\pi}{4}\left(j - \frac{1}{2}\right)\right]^{2/3} + E_0 \\ &= \hbar\Theta\xi + E_0\end{aligned} \tag{22}$$

The quantity ξ is the quasi-index. Equation (22) indicates that the electro-optic energy is obtained with use of the linear plot of $\hbar\omega$ with respect to ξ.

Figure 4 shows the photoreflectance spectrum of the undoped GaAs (*i*-GaAs)/*n*-type GaAs (*n*-GaAs) epitaxial structure at room temperature. The sample was grown on a (001)-oriented GaAs substrate by metal organic vapor phase epitaxy. The *i*-GaAs and *n*-GaAs layer thicknesses were 200 nm and 3.0 μm, respectively, and the doping concentration of the *n*-GaAs layer was 3×10^{18} cm^{-3}. In the photoreflectance measurement, the pump beam was the laser light with the photon energy of 2.33 eV chopped at the frequency of 2.0 kHz using an acoustic modulator. The pump-beam power was 2.0 mW. The probe beam was obtained from a tungsten-halogen lamp dispersed by a monochromator with a resolution of 0.5 nm. The probe-beam power was about 4 μW.

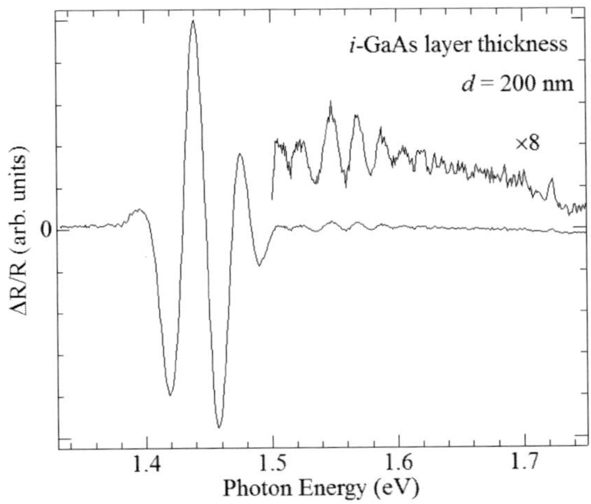

Figure 4. Photoreflectance spectrum of the *i*-GaAs/*n*-GaAs structure at room temperature. The oscillation patterns, the Franz-Keldysh oscillations, are located in the range from the fundamental transition energy of GaAs to the higher photon energy.

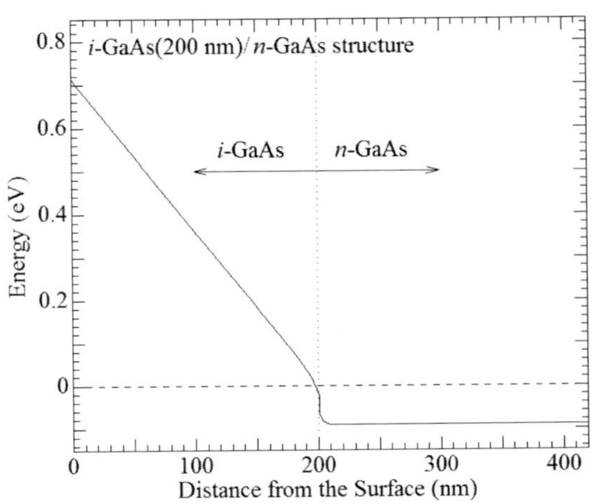

Figure 5. Calculated conduction-band energy in the *i*-GaAs/*n*-GaAs structure depicted as a function of distance from the surface.

In advance to discuss the photoreflectance spectrum, we present the evidence that the *i*-GaAs/*n*-GaAs structure has a uniform built-in electric field in the *i*-GaAs layer. Figure 5 shows the calculated conduction-band energy structure of the *i*-GaAs/ *n*-GaAs structure. The calculation was performed using a Boltzmann-Poisson method [13, 14]. As shown in Figure 5, the conduction band has a linear slope in the *i*-GaAs layer. This indicates the presence of the uniform electric field in the *i*-GaAs layer.

From now on, we evaluate the electric field in the *i*-GaAs layer with the analysis of the Franz-Keldysh oscillations in the photoreflectance spectrum. In the photoreflectance spectrum shown in Figure 4, the oscillation pattern is observed in the range from the fundamental transition energy of GaAs (1.425 eV) [15] to the higher photon energy (1.65 eV). The observed oscillation pattern is attributed to the Franz-Keldysh oscillations. The extremum photon energy positions of the observed Franz-Keldysh oscillations are plotted in Figure 6.

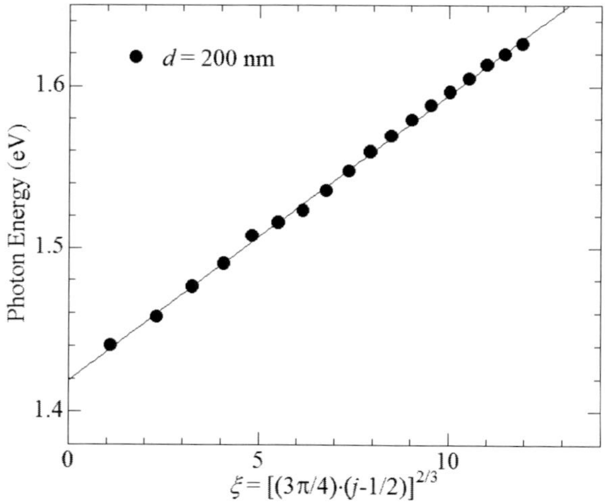

Figure 6. The extrema of the Franz-Keldysh oscillations plotted as a function of quasi-index ξ. The slope of the linear plot corresponds to the electro-optic energy $\hbar\Theta$.

The experimental data, which is denoted as the solid circles, are fitted to a linear function. The linear function well fits to the experimental data. The slope of the linear function corresponds to the electro-optic energy $\hbar\Theta$ as shown in Equation (22), so that the built-in electric field F is estimated from the value of $\hbar\Theta$. The value of F is estimated to be 28 kV/cm. In the estimation, the interband reduced mass was $0.556m_0$ [16]. Thus, the measurement of the

Franz-Keldysh oscillations is a useful way to evaluate the built-in electric field.

VI. Franz-Keldysh Oscillations in the Epitaxial Structures: In the Presence of the Effects of the Interference of the Probe Light Beam

We have treated the analysis of the Franz-Keldysh oscillations that are not influenced by the interference effect of the probe beam in Section III. In Section IV, we discuss the interference effects on the Franz-Keldysh oscillations resulting from multiple layer structures. Here, for the simplicity, we demonstrate the interference effects on the Franz-Keldysh oscillation resulting from the uniform built-in electric field in detail. Following this demonstration, we briefly describe the Franz-Keldysh oscillations that originate from the *p-n* junction involved in the epitaxial structures in the heterojunction bipolar transistors.

In order to clarify the interference effects, it is necessary to start from the amplitude reflectivity [17]. In the multiple layer structures with a total number of N and a substrate, the overall amplitude reflectivity $r(\hbar\omega)$ is written as

$$r(\hbar\omega) = r_{01}(\hbar\omega) + \sum_{n=1}^{N}\left(\prod_{j=1}^{n} t_{j-1,j}(\hbar\omega) t_{j,j-1}(\hbar\omega)\right) \times r_{n,n+1}(\hbar\omega) e^{i\delta_n(\hbar\omega)} \tag{23}$$

where $r_{n,n+1}$ and $t_{n,n+1}$ are the amplitude reflectivity and transmittivity at the n-th interface between the n-th and $(n+1)$-th layers. The layer indices of 0 and $N+1$ correspond to the ambient and the substrate, respectively. In Equation (23), we neglect the multiple reflections for the simplicity because of its relatively small contribution. The quantity $\delta_n(\hbar\omega)$ denotes the phase delay of the reflected probe beam between the n-th interface and the surface. The phase delay $\delta_n(\hbar\omega)$ is written as

$$\delta_n(\hbar\omega) = \sum_{m=1}^{n} 2k_m(\hbar\omega) d_m \tag{24}$$

In Equation (24), k_m and d_m are the wave vector of the probe beam with the energy of $\hbar\omega$ and the thickness in the m-th layer, respectively. The line-shape function $(\Delta R/R)_n$ of the Franz-Keldysh oscillations from the n-th interface is given by the combination of the following equations:

$$(\Delta R / R)_n \propto \mathrm{Re}[(C_n \Delta\varepsilon_n + C_{n+1}\Delta\varepsilon_{n+1})e^{i\delta_n(\hbar\omega)}] \tag{25}$$

Here $\Delta\varepsilon_n$ and C_n are the modulated dielectric function and Seraphin coefficients, respectively. In the derivation of the above equations, we utilized the relation between the overall modulated reflectance $\Delta R/R$ and the overall modulated amplitude reflectance $\Delta r/r$. In Equation (25), the quantity $\Delta\varepsilon_n$ is the modulated component of the dielectric function of the n-th layer. The quantity $\Delta\varepsilon_n$ is derived as shown in Equations (11) and (14). Consequently,

$$\frac{\Delta R}{R} \propto \frac{1}{4(\hbar\omega - E_0)(\hbar\omega)^2}\cos\left(\frac{4}{3}\left(\frac{\hbar\omega - E_0}{\hbar\Theta}\right)^{3/2} + \delta(\hbar\omega) + \frac{\pi}{2}\right) \tag{26}$$

From Equation (11), the j-th extremum of the Franz-Keldysh oscillations is given by

$$\hbar\omega_j = \hbar\Theta\xi_j + E_0 \tag{27}$$

where

$$\xi_j = \left[\frac{3\pi}{4}\left(j - \frac{\phi(\hbar\omega)}{\pi}\right)\right]^{2/3} \tag{28}$$

The function $\phi(\hbar\omega)$ is

$$\phi(\hbar\omega) = \frac{\pi}{2} + \delta(\hbar\omega) \tag{29}$$

Equation (28) has the interference term $\phi(\hbar\omega)$, which features the interference effect. The ξ_j which contains the interference term $\phi(\hbar\omega)$, is the extended quasi-index [18].

In order to indicate the appearance of the above discussion, we performed the following photoreflectance measurement. Figure 7 shows the photoreflectance spectrum from the i-GaAs/n-GaAs with the SiN_x cover film prepared by chemical vapor deposition (CVD). The i-GaAs/n-GaAs structure has the same structure described in Section III. The thickness of the SiN_x cover film was about 50 nm. The pump beam was the laser light with the photon energy of 2.33 eV at the frequency of 2.0 kHz using an acoustic modulator. The pump-beam power was 2.0 mW. The probe beam has a resolution of 0.5 nm. The spectra were obtained at the probe-beam power density of 30 $\mu W/cm^2$. In the spectrum, a large number of the Franz-Keldysh oscillations are observed.

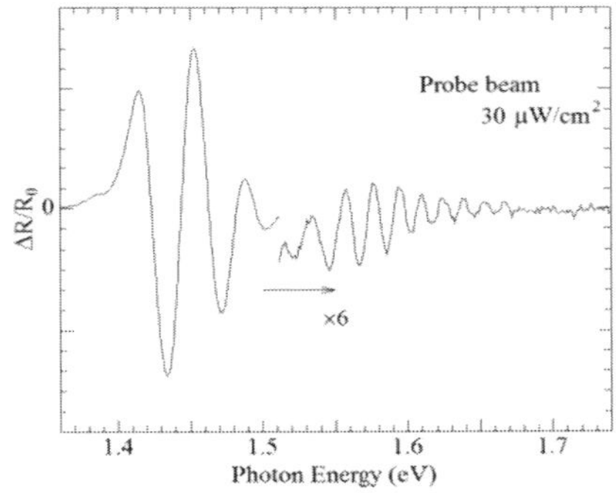

Figure 7. Photoreflectance spectrum in the SiN_x/*i*-GaAs/*n*-GaAs structure at room temperature. The SiN_x film thickness was 50 nm.

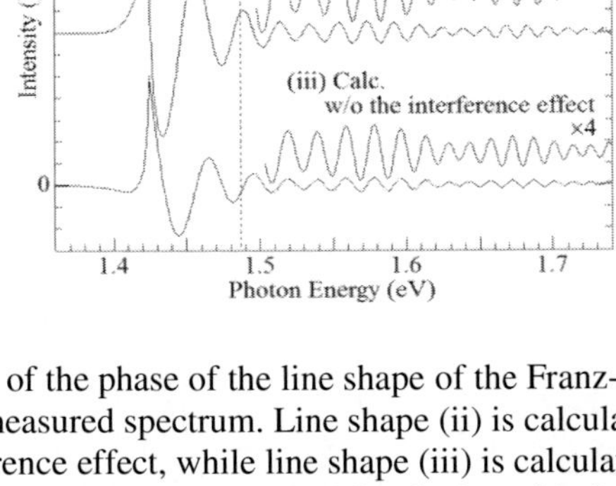

Figure 8. Comparison of the phase of the line shape of the Franz-Keldysh oscillations. Line shape (i) is the measured spectrum. Line shape (ii) is calculated with taking account of the interference effect, while line shape (iii) is calculated without taking account of the interference effect. The dashed line is the guide for the eyes for the comparison of the phase of the line shape. Reprinted with permission. Copyright (2004) American Institute of Physics.

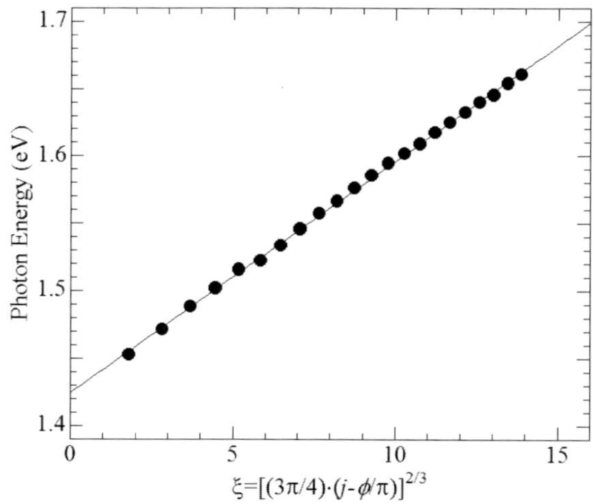

Figure 9. The extremum of the Franz-Keldysh oscillations plotted as a function of quasi index ξ. The slope of the linear plot corresponds to the electro-optic energy $\hbar\Theta$. Reprinted with permission. Copyright (2004) American Institute of Physics.

In order to confirm the reasonability of the line-shape analysis with use of Equation (26), Figure 8 shows the calculated line shapes of the photoreflectance spectra with/without taking account of the interference effects. Here, we focus our attention on the phase of the line shape. The dashed line is a guide for the comparison of the phase of the line shapes. The phase of the line shape, which is calculated without the interference effect, deviates the experimentally obtained spectrum. In contrast, the phase of the line shape, which is calculated with the interference effect, well agrees with the experimentally obtained spectrum. It is, therefore, confirmed that the line shape analysis with use of Equation (26) is appropriate. The extrema of the measured line shape of the Franz-Keldysh oscillations are plotted as a function of extended quasi-index. The line depicted as a function of quasi index is well fitted to the data for extrema shown in Figure 7. The obtained built-in electric field is 29.6 kV/cm, which is larger than the value obtained built-in electric field (28.0 kV/cm) for the sample without the cover film. This is attributed to the change in the surface condition that is induced during the deposition [18].

The above discussion is applicable to the layer of the epitaxial structure that has a uniform electric field. The built-in electric field in the *p-n* junction lineally changes along with the growth direction. In this cause, Equation (26) is modified as [19-21]:

$$\frac{\Delta R}{R} \propto \frac{1}{4(\hbar\omega - E_0)(\hbar\omega)^2} \cos\left(\frac{4}{3}\left(\frac{\hbar\omega - E_0}{\hbar\Theta}\right)^{3/2} + \delta(\hbar\omega)\right) \tag{30}$$

The difference between Equations (26) and (30) appears in the phase factor: the factor π/2 exists or not.

Figure 10 shows the photoreflectance spectra of the epitaxial structures for the $Al_xGa_{1-x}As/GaAs$-based heterojunction bipolar transistor structures with the 80-nm and 150-nm thick base layers. The photoreflectance spectra of the samples were measured at room temperature.

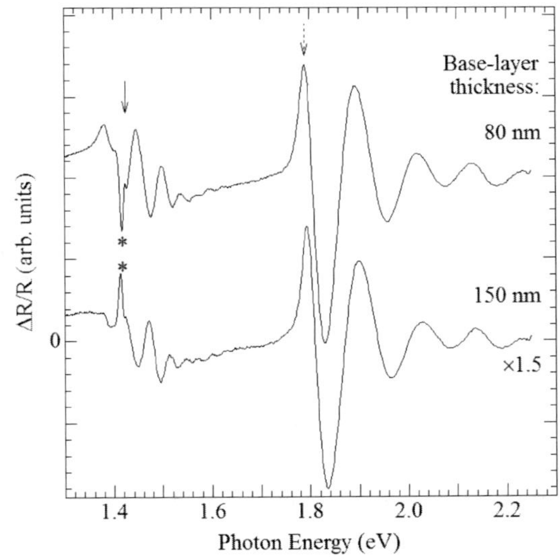

Figure 10. Photoreflectance spectra of the present $Al_{0.3}Ga_{0.7}As/GaAs$ heterojunction bipolar transistor epitaxial structures with the base-layer thickness of 80 and 150 nm at room temperature. The solid and dashed arrows indicate the positions of $E_{g,GaAs}$ and $E_{g,Al0.3Ga0.7As}$ respectively. The measurement was performed at room temperature. Reprinted with permission. Copyright (2004), American Institute of Physics.

The samples used for the measurements were two kinds of epitaxial structures for $Al_xGa_{1-x}As/GaAs$ heterojunction bipolar transistor grown on a 2°-off (001) GaAs substrate by metal organic vapor phase epitaxy. The two $Al_xGa_{1-x}As/GaAs$ heterojunction bipolar transistor epitaxial structures have a p^+-GaAs base layer. The base-layer thicknesses of the two samples are 80 and 150 nm. Except for the difference in the base-layer thickness, the epitaxial structures for the two $Al_xGa_{1-x}As/GaAs$ heterojunction bipolar transistors are the same. The emitter region of the $Al_xGa_{1-x}As$ GaAs heterojunction bipolar transistors are composed of three layers from the top: 30-nm thick compositionally graded n-$Al_yGa_{1-y}As(0 \leq y \leq 0.3)$, 115-nm thick n-$Al_{0.3}Ga_{0.7}As$, and 5-nm thick graded n-$Al_yGa_{1-y}As(0.1 \leq y \leq 0.3)$ layers. In contrast, the collector region has the following structure: the n^+-GaAs (15 nm) spacer layer adjacent to the base layer is followed by n^--GaAs (885 nm) collector, n^+-GaAs sub-collector and i-GaAs buffer layers, where the value in the parentheses denotes the individual layer thickness. The present $Al_xGa_{1-x}As/GaAs$ heterojunction bipolar transistor epitaxial structures have the following cap layers: n^+-$In_{0.5}Ga_{0.5}As$ (50 nm), graded n^+-In_yGa_{1-y} As ($0 \leq y \leq 0.5$, 50 nm) and n^+-GaAs (100 nm) emitter contact layers. In the photoreflectance measurement, the pump light was the laser light with the photon energy of 2.33 eV chopped at the frequency of 2 kHz with an acousto-optic modulator. The probe light was obtained from a tungsten-halogen lamp dispersed by a monochromator with a 1.2-nm resolution. The power density of the pump light was 240 mW/cm^2. The angle of the incidence of the probe light was 15°. The reflected probe light was detected with a Si photodiode.

In Figure 10, the solid and dashed arrows indicate the positions of the band-gap energies $E_{g,GaAs}$ of GaAs (1.424 eV) and of $Al_{0.3}Ga_{0.7}As$ (1.79 eV), respectively [15]. The Franz-Keldysh oscillations both from the GaAs-collector region and from the $Al_{0.3}Ga_{0.7}As$-emitter region are clearly observed in each spectrum. The oscillatory profile in the range above the $Al_{0.3}Ga_{0.7}As$ band-gap energy is composed of the Franz-Keldysh oscillations from the $Al_{0.3}Ga_{0.7}As$ and graded $Al_{1-y}Ga_yAs$ layers. The phase of the oscillatory profile does not change, irrespective of the base-layer thickness. The observed phenomenon means that the phase of the Franz-Keldysh oscillations is not shifted as long as the overlayer thicknesses are the same. The spectrum in the range between 1.4 and 1.7 eV, which is related to the photoreflectance signals from GaAs layers, consists of the three components: the oscillatory profile and the sharp structure marked by the asterisk at 1.42 eV are superposed on the gradually changing background signal. The phase of the background signal does not depend on the base-layer thickness. The layer producing the background signal is, therefore, assigned to the layers over the base region: the GaAs emitter contact or the graded n^+-$In_yGa_{1-y}As$ layers.

In contrast, the phases of the oscillatory profile and sharp structure depend on the base-layer thickness: both signals exhibit a phase inversion between the two photoreflectance spectra. This indicates that the oscillatory profile and sharp structure originate from the layer under the base region because the change in the phase reflects the difference in the overlayer thickness, according to Equation (30).

Next, we calculated the line shapes, which are compared with the measured spectra as shown in Figure 11, using Equation (30). In the calculation of the line shape, which originates from the collector-spacer region, the $(\Delta R/R)_{CS}$ of each sample. The used value of $E_{g,GaAs}$ is 1.424eV. The refractive indices applied to the calculation of $(\Delta R/R)_{CS}$ are listed in Table I [21]. In Figure 11, the calculated line shape of $(\Delta R/R)_{CS}$ is depicted above each measured spectrum as a dashed curve. Even in the calculation without any fitting parameters, the present line shape, which is calculated on the basis of Equation (30), well explains the phase inversion of the Franz-Keldysh oscillations between the two samples. This indicates the appropriateness of Equation (30). We note that the appropriateness of Equation (30) is also confirmed in the $In_{0.49}Ga_{0.51}P$/GaAs-based heterojunction bipolar transistor structures [22].

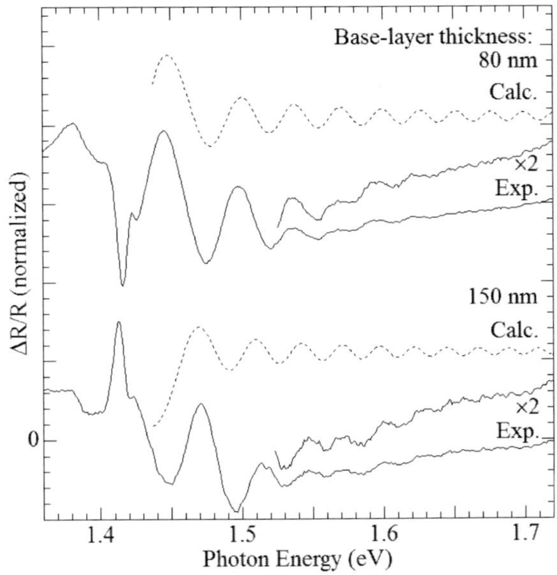

Figure 11. Calculated (dashed line) and measured (solid line) line shapes of the Franz-Keldysh oscillations from the collector-spacer junction in the $Al_{0.3}Ga_{0.7}As$ /GaAs heterojunction bipolar transistor structures with the base-layer thickness of 80 and 150 nm. For clarity, each spectrum is normalized. Reprinted with permission. Copyright (2004), American Institute of Physics.

V. FRANZ-KELDYSH OSCILLATIONS IN THE $AL_XGA_{1-X}N$/GAN HETERO STRUCTURES: HIGH SENSITIVITY TO THE SURFACE MORPHOLOGY

Photoreflectance spectroscopy has a nature of reflectance spectroscopy, so that photoreflectance measurement is applicable as a probe for the surface condition such as morphology. In Section V, we demonstrate the Franz-Keldysh oscillations in the $Al_xGa_{1-x}N$/GaN heterostructures and show the evidence of its high sensitivity to the surface morphorgy of the $Al_xGa_{1-x}N$ surface.

Figure 12 shows the images of three $Al_xGa_{1-x}N$/GaN heterostructure surfaces obtained with use of atomic force microscopy (AFM). In the present study, $Al_xGa_{1-x}N$/GaN heterostructures, samples A, B, and C, were grown on *c*-plane sapphire substrates by metal organic vapor phase epitaxy. Each sample has the same structure designed as follows from the top: a 27-nm-thick undoped $Al_{0.2}Ga_{0.8}N$ layer and a 2.2-μm-thick undoped GaN layer. In Figure 12, the AFM images of samples A, B, and C are shown in Figures12(a), 12(b), and 12(c), respectively. The samples have a difference in the $Al_{0.2}Ga_{0.8}N$-layer morphology. The $Al_{0.2}Ga_{0.8}N$ layer of sample A has a large number of cracks, which are not observed in samples B and C. The AFM image of sample B exhibits a lot of pits with a diameter of the order of 100 nm. In sample C, the pits are also observed; however, the density of the pits is much lower than that in sample B. In addition, the size of the pit in sample C is smaller than that in sample B. In usual [23, 24], each pit is formed on a specific threading

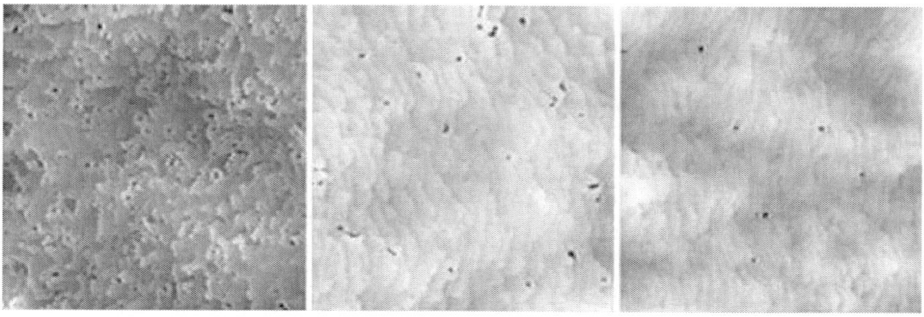

Figure 12. AFM images of the $Al_{0.2}Ga_{0.8}N$ surfaces in the area of 4.0 μm × 4.0 μm: (a) sample A, (b) sample B, and (c) sample C. Gray scale: corresponding height of 9.6 nm for (a), 8.6 nm for (b), and 5.3 nm for (c). With kind permission of The European Physical Journal (EPJ) (2007).

dislocation, and the specific dislocations hold a small fraction of the total dislocations. The specific dislocations gather a large number of impurities at its core, which leads to the suppression of the layer growth at the core area and the subsequent formation of the pits.

In order to clarify the sensitivity of the photoreflectance characterization to the surface morphorgy, the photoreflectance spectrum of each sample is shown in Figure 13. In the measurement of the photoreflectance spectra in Figure 13, the pump beam with a wavelength of 266 nm was provided from a diode-pumped Q-switched laser at the repetition rate of 1 kHz. The pulse width and average power density of the pump beam were 13 ns and 4.0 W/cm^2, respectively. The probe beam was monochromatic light with a spectral width of 1.2 nm.

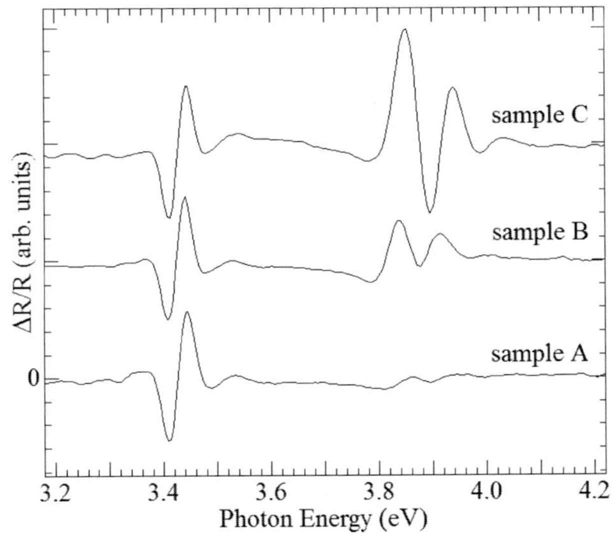

Figure 13. Photoreflectance spectra of samples A, B, and C at room temperature. With kind permission of The European Physical Journal (EPJ) (2007).

As shown in Figure 13, the photoreflectance spectra markedly depend on the $Al_{0.2}Ga_{0.8}N$-layer morphology. The signals at the photon energy of 3.42 eV, which is almost equal to the fundamental exciton-transition energy of GaN [25], are attributed to the photoreflectance signals from the GaN layer. Accordingly, the oscillations starting from 3.83 eV are assigned to the Franz-Keldysh oscillations from the $Al_{0.2}Ga_{0.8}N$ layer. The photoreflectance signals from the GaN layer have the same intensity in the three samples. In contrast, the amplitude of the Franz-Keldysh oscillations considerably decreases in order of samples C to A, which coincides with the lowering in the surface morphology that are observed with AFM images as shown in Figures 12(a)-12(c). This indicates that the cracks and pits reduce the amplitude of Franz-Keldysh oscillations; namely, the amplitude of the Franz-Keldysh oscillations is highly sensitive to the $Al_xGa_{1-x}N$-layer morphology. The presence of the imperfections such as cracks and pits leads to the following two mechanisms, which explain the Franz-Keldysh oscillations observed in samples A, B and C. One mechanism originates from lifetime broadening induced by the carrier scattering. The Franz-Keldysh oscillations are affected by crystallinity through a change in lifetime broadening [1, 7, 26]. The lowering of the crystal quality enhances the scattering and collisions of electrons, [26, 27] which causes the lifetime broadening of the Franz-Keldysh oscillations. Consequently, the presence of the imperfections, which remarkably shortens the carrier lifetime, enhances the lifetime broadening, and hence results in the reduction of the amplitude due to considerable oscillation damping. The other mechanism originates from the modulation-magnitude suppression. In the photoreflectance measurement, as mentioned in Section I, the modulation is connected with the built-in electric field. The imperfections act as traps and/or rapid recombination centers, and, as a result, reduce the total amount of the effective carriers contributing to the modulation. From the above-mentioned experimental results and discussion, the measurement of the Franz-Keldysh oscillations is applicable to evaluate the surface morphorgy characteristics.

In advance to finalizing Chapter V, it should be mentioned that the built-in electric field is also estimated from the Franz-Keldysh oscillations. Figure 14 shows the photon energies of the extrema, which are observed the Franz-Keldysh oscillations, as a function of quasi-index. The built-in electric fields, which are present the $Al_xGa_{1-x}N$ layer of samples B and C, are estimated to be 216 and 230 kV/cm, respectively, from the plot in Figure 14, in the same manner in Section III. In this estimation, the value of μ is 0.191 in units of free electron mass for the relevant transition [25]. The estimated built-in electric field is relatively high, which induces the two-dimensional electron gas at the $Al_xGa_{1-x}N$/GaN interface [28]. It is, therefore, the measurement of the Franz-Keldysh oscillations qualitatively provide information on the two dimension electron gas.

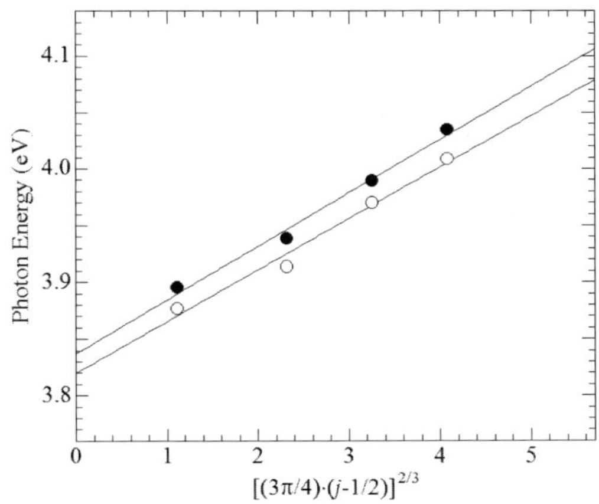

Figure 14. Positions of the extrema in the Franz-Keldysh oscillations plotted according to Equation (1). Lines: fitting results. Symbols: measured data. The solid circles are obtained for sample C, while the solid are for sample B. With kind permission of The European Physical Journal (EPJ) (2007).

VI. Estimation of the Built-in Electric Field Strength and Direction with Simultaneous Use of Photoreflectance Spectroscopy and Terahertz Spectroscopy

In the above Sections, the built-in electric field strength is estimated from the Franz-Keldysh oscillations; therefore, it is meaningful to evaluate the direction on the built-in electric field. In Section VI, it is demonstrated that the simultaneous use of the two measurements, photoreflectance spectroscopy and terahertz spectroscopy, effectively clarifies both the strength and the direction on the built-in electric field.

Figure 15 shows the terahertz waveform as a function of time delay in the $GaAs_{1-x}N_x$ epitaxial layer with x = 0.43% and the *i*-GaAs (Thickness: 200 nm)/*n*-GaAs (Thickness: 3 μm; doping concentration: 3×10^{18} cm^{-3}) epitaxial structure. The thickness of the $GaAs_{1-x}N_x$ epitaxial layer was 500 nm. The *i*-GaAs/*n*-GaAs epitaxial structure was used as a reference sample. As shown in Figure 5 of Section III, the conduction-band energy has a finite potential slope in the *i*-GaAs layer induced by the surface Fermi-level pinning, which produces a uniform built-in electric field of 28 kV/cm in the *i*-GaAs layer.

Such a relatively high electric field is sufficient to cause a surge current induced by the drift motion. Consequently, the electrons generated in the *i*-GaAs layer flow into the *n*-GaAs underlayer, while the holes generated in the *i*-GaAs layer flow into the *i*-GaAs surface side. The above-mentioned carrier flow enables the *i*-GaAs/*n*-GaAs structure to be a reference sample to check the polarity of the terahertz waves. The time-domain terahertz electromagnetic wave signals from the samples were measured with use of laser pulses with a duration time of 70 fs. The repetition rate of the laser pulse was 90 MHz. The pump beam was focused on the sample with the angle of incidence of 45°. The diameter of the spot on the sample surface was 100 μm. The emitted terahertz wave was collected with use of two off-axis parabolic mirrors, and was received by an optically gated bow tie antenna. The powers of the pump and gate beams were fixed to 40 and 4.0 mW, respectively. The wavelengths of the pump and gate beams had the same value of 800 nm. All the measurements were performed in the air at room temperature.

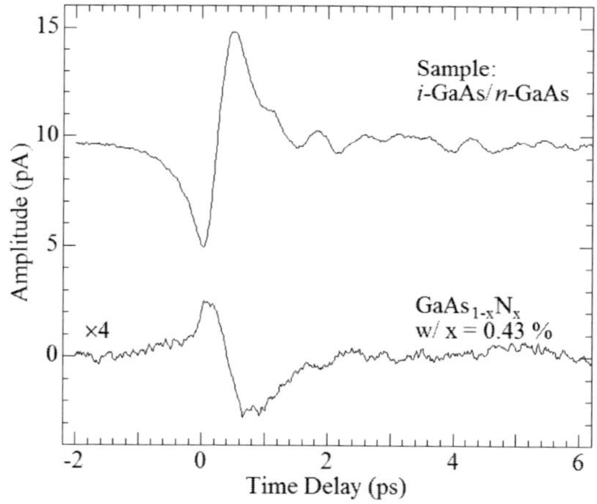

Figure 15. Amplitudes of the terahertz waveforms of the *i*-GaAs/*n*-GaAs and $GaAs_{1-x}N_x$ samples as a function of time delay at room temperature.

In Figure 15, the polarity of the terahertz waveform of the $GaAs_{1-x}N_x$ sample is inverted in comparison with that of the *i*-GaAs/*n*-GaAs sample. Accordingly, the $GaAs_{1-x}N_x$ sample has a downward potential bending: the electric filed direction is from the surface to inside.

In order to estimate of the built-in electric field strength, the photoreflectance spectrum of the $GaAs_{1-x}N_x$ sample was measured, which is shown in Figure 16. The position of the arrow labeled by E_- at 1.327 eV (E_+ at 1.747 eV) corresponds to the energy of the transition between the lower (upper) conduction subband and heavy hole and/or light hole bands at the Γ-point, and the position of the arrow labeled by $E_-+\Delta_0$ at 1.668 eV corresponds to the E_--subband—split-off-hole-band transition at the Γ-point, where the energies of E_-, E_+, and $E_-+\Delta_0$ are calculated on the basis of the band anticrossing model [29-31]. Consequently, the oscillation pattern starting from 1.3 eV is assigned to the Franz-Keldysh oscillations from the $GaAs_{1-x}N_x$ layer, while the oscillation pattern starting from 1.4 eV is assigned to the Franz-Keldysh oscillations from the GaAs buffer layer. In order to estimate the surface electric field, as shown in Figure 17, the extrema of the Franz-Keldysh oscillations from the $GaAs_{1-x}N_x$ sample is plotted as a function of quasi-index ξ_j. From the value of $\hbar\Theta$ the built-in electric field strength is estimated to be 24 kV/cm using an interband reduced mass of 0.0689 m_0 [32], where m_0 is the electron mass. In general, undoped GaAs crystals have a surface electric field of several kV/cm at most. The present results indicate that the incorporation of nitrogen enhances the surface electric field. Thus, it is concluded that the simultaneous use of photoreflectance spectroscopy and terahertz spectroscopy is effective to evaluate both the strength and the direction on the built-in electric field of the epitaxial structures.

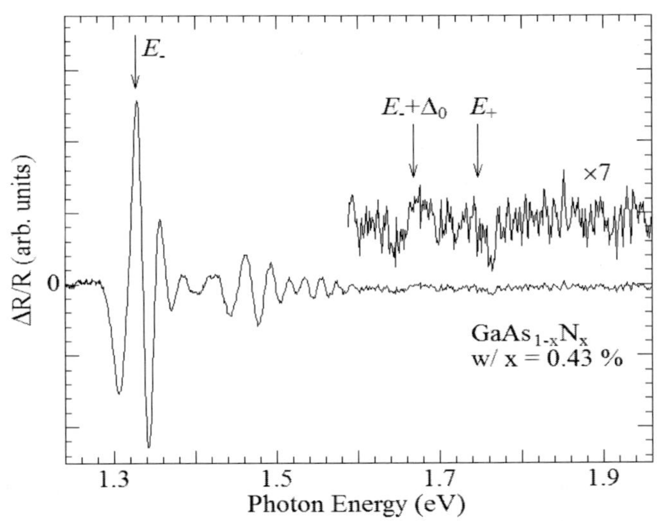

Figure 16. Photoreflectance spectrum of the $GaAs_{1-x}N_x$ sample at room temperature.

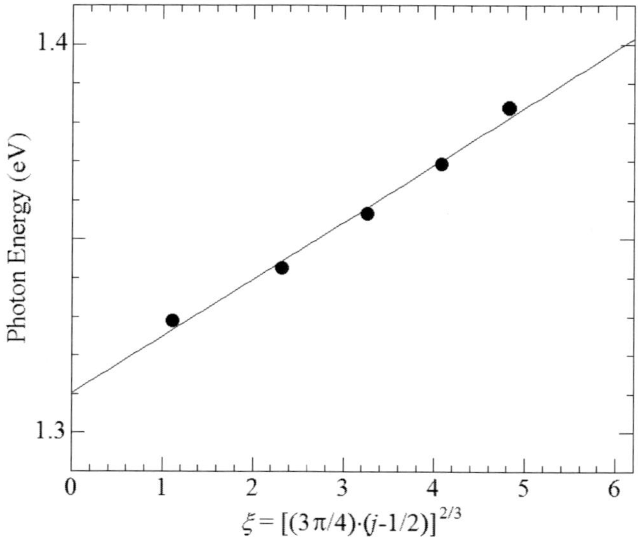

Figure 17. Linear plot of the extrema of the Fran-Keldysh oscillations from the $GaAs_{1-x}N_x$ sample as a function of the quasi-index .

VII. Application of the Franz-Keldysh Oscillation to the Structure Desing of the Terahertz Wave Emitters

As discussed the former Sections, the Franz-Keldysh oscillations are considerably sensitive to the surface and/or built-in electric field. Recent progress in femtosecond-pulse laser technology enables cutting-edge terahertz spectroscopy [33]. In the research and developmental field of terahertz spectroscopy, the design guide line is still investigated. Here, we demonstrate that the measurement of the Franz-Keldysh oscillations is applicable to the optimization of the simple terahertz emitters: the *i*-GaAs/*n*-GaAs emission from the coherent longitudinal optical (LO) phonons [34].

Before proceeding in the experimental results and discussion, we present the background of the intensive research for the terahertz emission from the coherent LO phonons. We also briefly describe the terahertz emission of the coherent LO phonons.

The terahertz emission from coherent LO has been attracting much attention since the development of monochromatic terahertz emitters is an important issue in terahertz spectroscopy. In general, however, the intensity of the terahertz emission from coherent optical phonons is weak in bulk semiconductor crystals [35]. As a solution of this issue, the application of the multiple quantum wells was proposed [36, 37]. In the multiple quantum wells, the terahertz emission from the coherent LO phonon is enhanced in the case where the fundamental heavy-hole and light-hole exciton energy spacing is equal to the LO phonon frequency. In addition, the photon energy of the pump beam should be tuned to the centre energy between the heavy-hole and light-hole exciton energy spacing: The quantum interference between the heavy-hole and light-hole excitons is a driving force for the coherent LO phonon. The above method requires the strict sample growth and limits the photon energy from the coherent LO phonons. The coherent longitudinal optical (LO) phonons are also induced by the surge current of photogenerated carriers that changes a potential structure in equilibrium. The generation mechanism of the coherent phonons indicates that the substantial trigger for driving the coherent LO phonons is not the illumination of the pump beam but the resultant potential modulation. The presence of the surface electric field (**E**) results in the enhancement of the polarization (**P**) of the LO phonon along the direction of the electric field, which is well established as the following equation: $\mathbf{P} = [\chi]\mathbf{E}$, where $[\chi]$ is an electric susceptibility tensor [17]. The enhancement of the polarization connects with the enlargement of the initial displacement of the LO phonon. The initial displacement is released by an instantaneous change of the electric field by the screening effect due to photogenerated carriers, which launches the coherent oscillation of the LO phonon, *i.e.*, the coherent LO phonon. This phenomenon leads to the oscillation of the LO-phonon polarization producing the terahertz wave. It is noted that the enlargement of the initial displacement results in the enhancement of the amplitude of the coherent LO phonon. Consequently, the presence of the built-in electric field leads to the enhancement of the terahertz-wave intensity from the coherent LO phonon. This phenomenon provides an additional degree of freedom to the strategy for enhancing the monochromatic terahertz wave from the coherent LO phonons, which connects with the design strategy for the realization the *i*-GaAs/*i*-GaAs terahertz emitters that produce the terahertz wave from the coherent LO phonons. It should be noted that the translational symmetry is lowered in the surface region because of the presence of the relatively high electric field. This phenomenon evidently is described as Equation (1). This is advantageous for the realization of the terahertz emission from the coherent LO phonons.

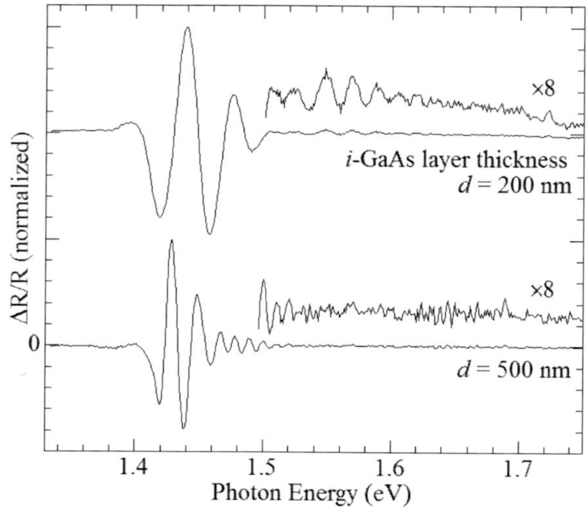

Figure 18. Photoreflectance spectra of the *i*-GaAs(200 nm)/*n*-GaAs and *i*-GaAs(500 nm)/*n*-GaAs samples at room temperature. The *i*-GaAs layer thickness *d* is denoted in the paresis. Copyright (2010) The Japan Society of Applied Physics.

The present samples were *i*-GaAs(*d* nm)/*n*-GaAs structures grown on semi-insulating (001)-oriented GaAs substrates by metal organic vapor phase epitaxy. The layer thickness and doping concentration of the *n*-GaAs layer were 3 μm and 3×10^{18} cm^{-3}, respectively. The values of *d* were 200, 500, 800, and 1200 nm. In advance to the terahertz-wave measurement, we measured the photoreflectance spectrum of each sample at room temperature in order to evaluate the relation between the built-in electric filed and *d*. In the photoreflectance measurement, the pump beam was the laser light with the photon energy of 1.96 eV chopped at the frequency of 630 Hz. The pump-beam power was 2.0 mW. The probe beam was obtained from a tungsten-halogen lamp dispersed by a monochromator with a resolution of 0.5 nm. The probe-beam power was about 4 μW.

Figure 18 shows the photoreflectance spectra of the *i*-GaAs(200 nm)/*n*-GaAs and *i*-GaAs(500 nm)/*n*-GaAs samples. In the photoreflectance spectra, the oscillation patterns, which originate from the Franz-Keldysh oscillations, are observed. In order to estimate the built-in electric field, as shown in Figure 18, the extrema of the Franz-Keldysh oscillations from the *i*-GaAs layer are plotted as a function of the quasi-index $\xi \equiv [(3\pi/4)\cdot(j-1/2)]^{2/3}$, where j denotes the index of each extremum numbered from the fundamental transition energy position. The slope of the solid line is the electro-optic energy $\hbar\Theta$ given by $(e^2\hbar^2F^2/2\mu)^{1/3}$, where F and μ are the built-in electric field and interband reduced effective mass, respectively. From the slope of the solid line shown in Figure 19, it is evident that the built-in electric field of the *i*-GaAs(200 nm)/*n*-GaAs sample is stronger than that of the *i*-GaAs(500 nm)/*n*-GaAs sample. It is, therefore, confirmed that the built-in electric field of the *i*-GaAs(*d* nm)/*n*-GaAs sample is enhanced by decreasing *d*. The built-in electric fields of the *i*-GaAs(*d* nm)/*n*-GaAs samples are estimated to be 28, 12, 8.2, and 6.1 kV/cm for d = 200, 500, 800, and 1200 nm, respectively, using the relation between the electric field and the electro-optic coefficient: $\hbar\Theta = (e^2\hbar^2F^2/2\mu)^{1/3}$.

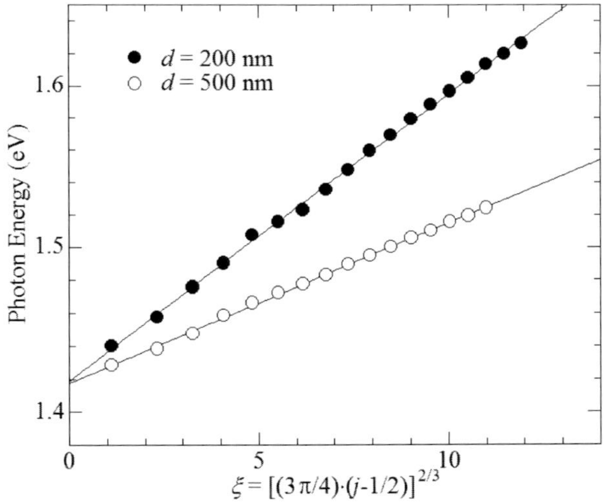

Figure 19. Plot of the extrema of the Franz-Keldysh oscillations from the *i*-GaAs(200 nm)/*n*-GaAs and *i*-GaAs(500 nm)/*n*-GaAs samples as a function of quasi-index ξ. The closed and open circles indicate the linear plots of the Franz-Keldysh-oscillations extrema of the *i*-GaAs(200 nm)/*n*-GaAs and *i*-GaAs(500 nm)/*n*-GaAs samples, respectively. Copyright (2010) The Japan Society of Applied Physics.

Next, we performed the time-domain signals of the terahertz waves, which are shown in Figure 20. The signals of the time-domain terahertz wave from the samples were measured using laser pulses with a duration time of about 70 fs. The repetition rate of the laser pulse was about 90 MHz. The pump beam was focused on the sample with the angle of incidence of 45°. The diameter of the spot on the sample surface was about 100 μm. The emitted terahertz wave from the sample was collected with use of two off-axis parabolic mirror and was detected by an optically gated dipole antenna with a gap of 6.0 μm formed on a low-temperature-grown GaAs epilayer. In Figure 20, the monocycle signal, which originates from the surge current of photogenerated carriers, appears around the time delay of 0 ps. The monocycle signal is followed by the relatively long oscillation pattern with a period of 112 fs. The period of 112 fs corresponds to the frequency of the coherent GaAs LO phonons; namely, the terahertz emission of the coherent GaAs LO phonons using the *i*-GaAs(*d* nm)/*n*-GaAs structures.

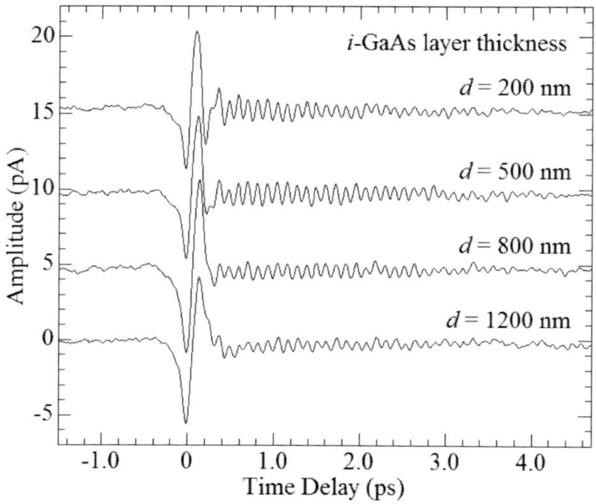

Figure 20. Terahertz waveforms of the *i*-GaAs(*d* nm)/*n*-GaAs structures. Copyright (2010) The Japan Society of Applied Physics.

In order to analyze the terahertz waveform, we performed the Fourier transform. The Fourier power spectra are shown in Figure 21. The bands, which appear in the range from 0 to 6.0 THz, result from the surge current of the photogenerated carriers that corresponds to the monocycle signal shown in Figure 20. The band peaking at 8.8 THz originates from the coherent LO phonon. It should be mentioned that the peak intensity of the coherent GaAs LO phonons in the *i*-GaAs(200 nm)/*n*-GaAs and *i*-GaAs(500 nm)/*n*-GaAs structures is larger than the peak intensity of the monocycle signal. The intensity of the coherent LO phonon is relatively high in comparison with the intensity of the monocycle signal, so that the present strategy, which uses the photoreflectance measurement of the Franz-Keldysh oscillations, is well reasonable; namely, the clarification of the built-in electric field is effective in designing the terahertz emitters. Although the electric field strength is enhanced by the increase in the *i*-GaAs layer thickness *d*, the peak intensity of the coherent LO phonon, which is in the *i*-GaAs(500 nm)/*n*-GaAs structure, is larger than that of the *i*-GaAs(200 nm)/*n*-GaAs structures. This phenomenon is attributed to the trade-off effect: the volume and built-in electric field.

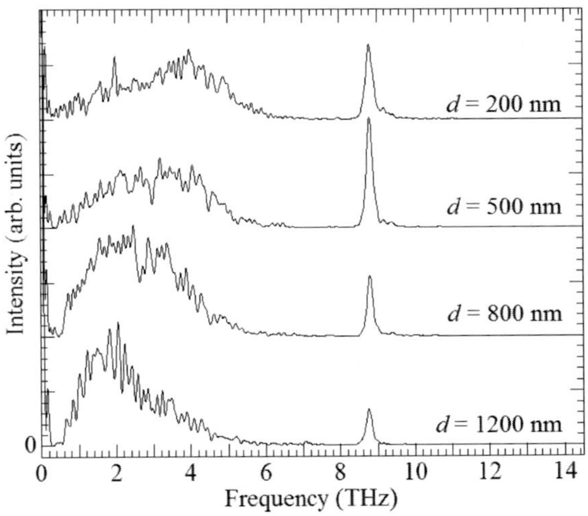

Figure 21. Fourier power spectra of the terahertz waveforms shown in Figure 20.

CONCLUSION

In the present chapter, we have demonstrated the effectiveness of photoreflectance spectroscopy of the Franz-Keldysh oscillations in order to evaluate the semiconductor heterostructures. In the present chapter, we have focused our attention on the interference effect on the semiconductor heterostructures. We have presented the theories and compared the experimental results with the calculation. It has been clarified that, taking account of the interference effects, the phase change of the Franz-Keldysh oscillations is well explained. We have applied the photoreflectance measurement to the evaluation of the surface homology of the $Al_xGa_{1-x}N/GaN$ heterostructures. The Franz-Keldysh oscillations, which originate from the $Al_xGa_{1-x}N/GaN$ layer, are drastically reduced in the enhancement of the amplitude of the Franz-Keldysh oscillations in order of lowering surface morphology. We have proposed built-in electric field strength and direction with simultaneous use of photoreflectance spectroscopy and terahertz spectroscopy. It has been found that the above-mentioned simultaneous use is effective to clarify the characteristics of the built-in electric field. Thus, we conclude that, for the evaluation of the heterostructures, photoreflectance spectroscopy of the Franz-Keldysh oscillations markedly and significantly characterizes the heterostructures; namely, the measurement of the Franz-Keldysh oscillations considerably enhances the research and development of the semiconductor heterostructures.

REFERENCES

[1] M. Cardona, *Modulation spectroscopy*, Academic Press, New York (1969).

[2] X. Yin, F.H. Pollak, L. Pawlowicz, T. O'Neill, and M. Hafizi, *Appl. Phys. Lett.* 56, 1278 (1990).

[3] X. Yin, F.H. Pollak, L. Pawlowicz, T. J. O'Neill, and M.E. Hafizi, *Proc. SPIE* 1286, 404 (1990).

[4] W. Franz, *Z. Naturforschung* 13a, 484 (1958).

[5] L.V. Keldysh, *J. Exptl. Theoret. Phys.* (USSR) 33, 994 (1957), translation: *Soviet Physics JETP* 6, 763 (1958).

[6] L. V. Keldysh, *J. Exptl. Theoret. Phys.* (USSR) 47, (1964), translation: *Soviet Physics JETP* 20, 1307 (1965).

[7] D.E. Aspnes, *Phys. Rev.* 153, 972 (1967).

[8] D.E. Aspnes, *Phys. Rev. B* 10, 4228 (1974).

[9] D.E. Aspnes and A.A. Studna, *Phys. Rev. B* 7, 4605 (1973).

[10] F.H. Pollak and H. Shen, *Mater. Sci. Eng. R* 10, 275 (1993).

[11] P. Yu and M. Cardona, *Fundamentals of Semiconductors: Physics and Materials Properties* (4th edition), Springer (2010), pp.322.

[12] B.O. Seraphin and N. Bottka, *Phys. Rev.* 145, 628 (1966).

[13] P.A. Basore, *IEEE Trans. Electron. Devices* 37, 337 (1990).

[14] D.A. Clugston and P.A. Basore, Conf. Rec. 26th IEEE Photovoltaic Specialists Conference, (IEEE, Piscataway, NJ, 1997) p. 207.

[15] O. Madelung (Ed.), *Semiconductors –Basic Data,* Springer, Berlin (1996).

[16] D.F. Nelson, R.C. Millar, C.W. Tu, and S.K. Sputz, *Phys. Rev. B* 36, 8063 (1987).

[17] D. J. Griffiths, *Introduction to Electrodynamics* (4th edition), Pearson, Boston (2013).

[18] H. Takeuchi, Y. Kamo, Y. Yamamoto, T. Oku, M. Totsuka, and M. Nakayama, *J. Appl. Phys.* 97, 063708 (2005).

[19] H. Takeuchi, Y. Yamamoto, R. Hattori, T. Ishikawa, and M. Nakayama, *IEICE Transactions on Electronics* E86-C, 2015 (2003).

[20] H. Takeuchi, Y. Yamamoto, R. Hattori, T. Ishikawa, and M. Nakayama, *Jpn. J. Appl. Phys.* 42, 6772 (2003).

[21] H. Takeuchi, Y. Yamamoto, and M. Nakayama, *J. Appl. Phys.* 96, 1967 (2004).

[22] H. Takeuchi, Y. Yamamoto, R. Hattori, T. Ishikawa, and M. Nakayama, *Physica E* 21, 693 (2004).

[23] Z. Liliental-Weber, Y. Chen, S. Ruvimov, and J. Washburn, *Phys. Rev. Lett.* 79, 2835 (1997).

[24] Y. Chen, T. Takeuchi, H. Amano, I. Akasaki, N. Yamada, Y. Kaneko, S.Y. Wang, *Appl. Phys. Lett.* 72, 710 (1998).

[25] Vurgaftman, J. R. Meyer, and L. R. Ram-Mohan, *J. Appl. Phys.* 89, 5815 (2001).

[26] C.F. Klingshirn, *Semiconductor Optics*, Springer, Berlin (4th edition) (2012)

[27] R. Enderlein, Phys. Status Solidi 20, 295 (1967).

[28] O. Ambacher, J. Smart, J.R. Shealy, N.G. Weimann, K. Chu, M. Murphy,W.J. Schaff, L.F. Eastman, R. Dimitrov, L. Wittmer, M. Stutzmann, W. Rieger, J. Hilsenbeck, *J. Appl. Phys.* 85, 3222 (1999).

[29] W. Shan, W. Walukiewics, J. W. Ager III, E. E. Haller, J. F.Geisz, D. J. Friedman, J. M. Olson, and S. R. Kurz, *Phys. Rev. Lett.* 82, 1221 (1999).

[30] W. Shan, W. Walukiewics, K. M. Yu, J. W. Ager III, E. E. Haller, J. F.Geisz, D. J. Friedman, J. M. Olson, S. R. Kurz, C. Nauka, *Phys. Rev. B* 62, 4211 (2000).

[31] W. Walukiewics, K. Alberi, J. Wu, W. Shan, K. M. Yu, and J. W. Ager III, in "Dilute III-V Nitride Semiconductors and Material Systems" ed. by A. Erol (Springer, Berlin 2008) pp.65-89.

[32] The calculations for the quantities were performed with use of the equations and parameters in Reference 31, pp.128-129.

[33] K. Sakai (ed.) *Terahertz Optoelectronics*, Springer, Berlin (2005) and relevant references therein.

[34] H. Takeuchi, J. Yanagisawa, S. Tsuruta, H. Yamada, M. Hata, and M. Nakayama, *Jpn. J. Appl. Phys.* 49, 082001 (2010).

[35] P. Gu and M. Tani in *Terahertz Optoelectronics*, ed. K. Sakai, Springer, Berlin (2005) pp.63-97 and relevant references therein.

[36] K. Mizoguchi, T. Furuichi, O. Kojima, M. Nakayama, S. Saito, A. Syouji, and K. Sakai, *Appl. Phys. Lett.* 87, 093102 (2005).
[37] M. Nakayama S. Ito, K. Mizoguchi, S. Saito, and K. Sakai, *Appl. Phys. Express* 1, 012004 (2008).

In: Optical Spectroscopy
Editor: Nicolae Tomozeiu

ISBN: 978-1-63321-197-1

Chapter 4

IR ATTENUATED TOTAL REFLECTION: A TOOL TO INVESTIGATE LIQUID PENETRATION IN PAPER: THEORETICAL CONSIDERATIONS

Nicolae Tomozeiu*
A&PS, R&D Océ Technologies, Venlo, The Netherlands

ABSTRACT

Liquid penetration into porous media has been the subject of important research with very good scientific achievements and technological implementations. However, developments and new tools to investigate the penetration of multiphase fluids into paper like porous media are still needed. New trends in printing technologies, where the high print quality must be combined with environmentally friendly requirements motivate us to focus on this subject.

Using Attenuated Total Reflection in infrared region (IR-ATR spectroscopy), the water penetration from thin liquid layers of aqueous solutions in paper is investigated. The structure of paper, revealed by high resolution SEM spectra, is porous consisting of cellulose fibers that interact via hydrogen bonds and fillers. A thin film of the liquid deposited on the surface of the ATR crystal is covered by this porous structure and the ATR spectrum recorded periodically reveals valuable information

* E-mail address: Nicolae.Tomozeiu@Oce.com.

about the liquid transport. The evolution in time of the signal assigned to water phononic vibration is representative for liquid penetration in paper. The influence of the liquid properties on the penetration behavior is revealed by studying both water and glycerol as liquids interacting with the same paper under the same conditions.

Keywords: ATR spectroscopy, porous paper, liquid penetration in porous media

1. INTRODUCTION

The penetration of liquid in porous materials has been the subject of many publications in the last decades, both theoretical understanding and sustainable industrial applications being the driving forces for these studies. Among porous materials, the paper utilized in writing and/or printing occupies a special place; understanding the transport of a liquid with well defined physical properties (e.g., water based ink) through/into it is important. The internal structure of paper is complicated, presenting anisotropy and degrees of heterogeneity due to various cellulose fibers, fillers ($CaCO_3$, clay, etc.) and pores in a large variety in dimension and shape [1]. This makes difficult the modeling of liquid penetration in such a porous material. Moreover, this problem is complicated by fiber swelling caused by water in both the liquid and vapor phases. As an example of a proposed model to describe the liquid penetration in paper we note the work of Verhoeff et al. [2] where there are considered two mechanisms for the liquid transport: a) capillary flow through the pores, and b) diffusional flow through fibers. These processes are so fast that they could not be separated.

Various investigation methods to reveal the liquid penetration in porous paper have been used: Bristow absorption tester [3], Scanning electron microscope [4], ultrasonic techniques [5], Near-Infrared spectroscopy [6], pulsed field gradient (PFG) nuclear magnetic resonance (NMR) [7], electrical capacity measurements [8], etc. In this work we present another spectroscopic method to investigate the liquid penetration in porous paper: attenuated total reflection in infrared domain. The first part is dedicated to designing the theoretical framework for this method and the proof of principle is made considering one experimental set of data obtained by measuring water penetration in plain paper. In the second part [9], dedicated to the experimental work, we propose two methods of investigation using the IR-ATR spectroscopy and we analyze the results using the model proposed in the first part.

After an introduction to the ATR measurement method, the signals assigned to molecular vibration of water and paper are formulated considering the ATR signal description. The liquid penetration in porous paper is the subject of section 4, where a brief review of the existing models is made. Then, the mathematical description of liquid penetration length from the measured signal is shown. The validation of the proposed model is made by analyzing the experimental data in section 5.

2. On the ATR Method: Theoretical Considerations

When an electromagnetic wave is incident on the interface between two different optical media characterized by the refractive index n_1 and n_2, respectively, the phenomena of reflection and refraction occur. The theory that describes these physical phenomena is given in text books of light propagation [10, 11]. For a good understanding of the fundamentals of the ATR method, let's consider an IR beam that comes from the medium n_1, to the interface, as shown in Figure 1.

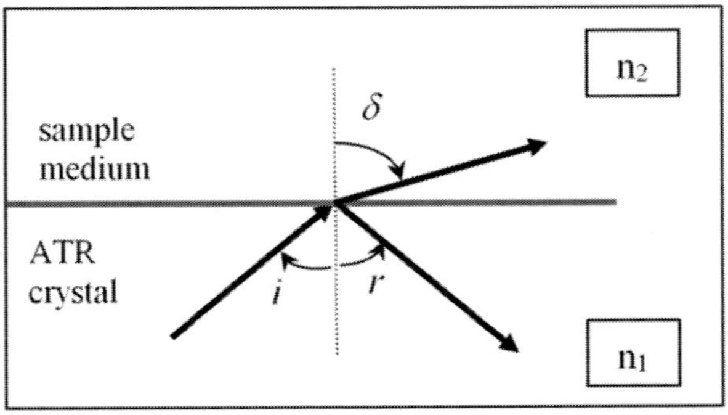

Figure 1. Simple sketch showing the reflection and refraction phenomena of the optical waves at the interface between two media with refractive index n_1 and n_2 respectively.

The necessary conditions for ATR are: $n_1>n_2$, and the incident optical beam comes from the denser medium (higher refractive index). Considering the optical law of refraction[†] in this condition, the refractive angle, δ, is larger than the incident angle, i, according to the relation (1):

$$\sin(\delta) = \frac{n_1}{n_2} \cdot \sin(i) \tag{1}$$

This is always satisfied as long as $n_1 > n_2$. Moreover, the refractive angle δ can be 90° if the incident angle equals the so called critical angle, *l*, value:

$$l = \arcsin\left(\frac{n_1}{n_2}\right) \tag{2}$$

For all values of the incident angle larger than the critical angle, $i > l$, the refracted beam is "returned" in the incident medium; from a physics point of view, this is the total reflection phenomenon.

Getting more into detail, for total internal reflection, the electrical field amplitude has a sinusoidal variation with the distance from the surface, in the medium n_1 (in Figure 1 is the "ATR crystal" medium, which is characterized by a larger mass density) and exponentially decreases in the rarer medium n_2 (the sample medium in Figure 1). By choosing the right incident angle and/or the relation between n_1 and n_2, the amplitude of the electrical field in medium n_2 can be maximized at the surface and a depth of penetration can be calculated:

- the electrical field assigned to this wave in sample decreases exponentially with the distance, z, from the sample/crystal interface:

$$E = E_0 \cdot \exp\left(-\frac{z}{\delta_p}\right) \tag{3}$$

where E_0 is the intensity of the electrical field at the interface (z=0) and δ_p is the depth of penetration of the evanescent electrical field;

- the latter is given by:

[†] The so-called Snellius-Descarte's law of refraction.

$$\delta_p = \frac{\lambda}{2\pi \cdot \sqrt{n_1^2 \cdot \sin^2\theta - n_2^2}} \tag{4}$$

with λ – the wavelength of interest, θ is the crystal angle.

Intuitively δ_p represents the place measured in the propagation direction, where the electrical field of the electromagnetic wave decreases "e" times in comparison with its value at the surface. The real depth investigated by the ATR beam was found larger than δ_p value. This was reported by Mirabella [12] who found the actual sampling depth to be approximately $3\delta_{p.}$

The depth of penetration of the radiation into the analyzed sample is important to be as large as possible. In this way the measured signal is more accurate. Considering relation (4), the ATR crystal characterized by its refractive index, n_1, and the geometrical details (the crystal angle θ**)** is the element of the experiment that can be easily chosen. The refractive index for some recommended materials as ATR crystals is given in Table 1. Considering the second medium a hypothetical material sample with n_2=1.5, the critical angle values for these crystals/sample interfaces can be calculated accordingly to relation (2). They are given in the same Table 1 and we can conclude that Ge and Si are the materials most recommended.

Having the material for the ATR crystal, the geometrical parameters as the angle θ, the thickness Δ and the length B of the crystal are of importance in designing a good experiment.

Table 1. Refractive index of some materials used as ATR crystals

Material crystal	**n_1**	**l (degrees)**
Ge	4	22.04
Si	3.42	26.01
GaAs	3.14	28.54
CdTe	2.65	34.47
ZnSe	2.42	38.30
ZnS	2.24	42.04
AgBr	2.2	42.99

These geometrical parameters are related between them via the number of reflections, N, that the beam experiences at the interface of the ATR crystal with the investigated sample (see the Rel. (5)). Sending the incident beam at the normal incidence on the inclined surface that makes the angle θ with the horizontal surface of the crystal, it will reach the interface under an angle of incidence i=θ. The length B is:

$$B = (N + 2) \cdot \Delta \cdot \tan(\theta). \qquad (5)$$

In other words, the number of reflections on the ATR crystal/sample interface is well defined by the geometry of the measurement setup. This is very important because each reflection acts as an amplifier for the sample fingerprint contained in the output beam analyzed by the detector.

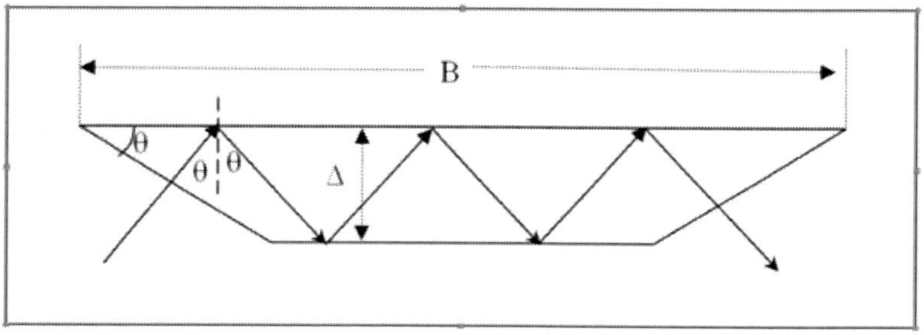

Figure 2. The beam path inside an ATR crystal considering the angle, thickness and the length of the crystal.

The electrical field of the IR wave interacts with the electrical dipoles that exist in the volume defined by the "sampling depth" and the ATR crystal area, and excites their vibration modes. Energy will be transferred to these vibration modes and bands of absorption will appear in the IR spectrum. These absorption lines/bands are specific to the molecular species with electrical dipoles in the investigated volume. Besides their position in spectrum (wave-numbers), the intensity of these absorptions is important and it can be related to the species concentration. For this, let's consider that the IR radiation which has its intensity proportional to the E^2 (with E given by relation (3)) interacts with atoms/dipoles in concentration "c" in the investigated sample. This interaction is characterized by the energy absorption, revealed by the absorption coefficient γ. Under these circumstances an incident radiation with intensity I_o will yield an ATR signal:

$$\Im_o = \chi \cdot N \cdot \int_0^L c \cdot \left(\gamma \cdot E^2\right) \cdot dz$$

or, considering the relation (3),

$$\Im_o = \chi \cdot N \cdot \int_0^L c \cdot \gamma \cdot E_0^2 \cdot \exp\left(-\frac{2z}{\delta_p}\right) \cdot dz \quad (6)$$

where χ is a constant that characterizes the measurement instrument, z is the distance on the perpendicular direction to the ATR crystal into the investigated sample. It has to be noted that here $L \sim 3\delta_p$, and the multiplicative constant of the instrument can be normalized by using a calibration sample.

Another important issue in this type of measurements is the air gap between the ATR crystal and the sample to be measured placed on it. Using ATR data measurements/analyses and spectra simulation it was found that the ATR signal exponentially decreases with the thickness of the air gap:

$$\Im \propto \exp\left(-\frac{d_{air}}{\beta}\right)$$

where β=403.8 nm for an investigated peak absorption at 825.6 cm^{-1}.

Most of the modern ATR setups have special accessories to avoid the existence of such an air gap between the crystal and the sample. When measuring rough sample, the experimentalist has to consider this effect and to avoid it in measurements.

3. Modeling the ATR Spectra

3.1. ATR Signals Assigned to Water and Paper

Let's consider an IR-ATR experiment where the sample to be investigated is pure water. A sketch of this experiment is shown in Figure 3, where the case of one reflection of the IR beam at the ATR crystal/water interface (N=1) is considered. As long as the thickness of the sample placed on the ATR crystal is larger than the depth $L \sim 3\delta_p$ and the concentration of the studied species is constant, the ATR signal is constant. When the thickness decreases and/or the dipoles' concentration decreases, the ATR signal diminishes.

The IR spectrum of water measured within such an experiment is shown in Figure 4. The absorption peak centered at 1639cm^{-1} (1639.2cm^{-1} in Figure

4) is assigned to water molecule vibrations, while the large band in the region of 3100 – 3600cm^{-1} describes the OH entities. In most of the situations, when the evolution in time of water molecules in the investigated volume is studied, the peak centered at 1639cm^{-1} is followed. In the presence of other types of molecules that contain the OH group (e.g., glycerol, or cellulose), the broad absorption band at 3300cm^{-1} must be studied in correlation with other absorption peaks assigned to these molecules; otherwise the information obtained can be non-accurate. Therefore, in this work, the representative absorption feature for water is the peak centered at 1639cm^{-1} (or 6097.5nm, in wavelength). The refractive index corresponding to this wavelength is 1.327. These data are needed to estimate the depth of penetration for this wavelength in water.

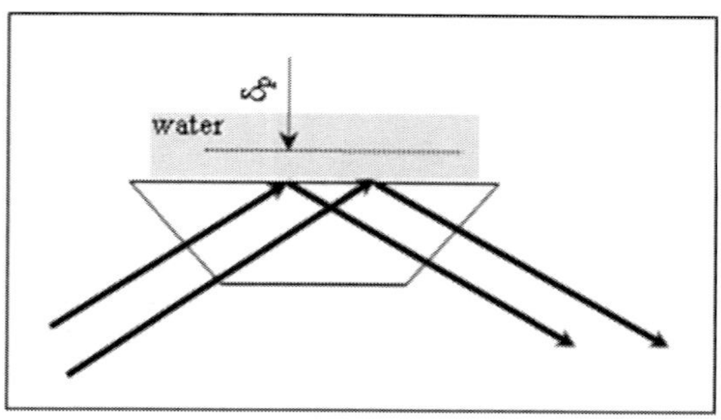

Figure 3. The sketch of an ATR experiment with N=1 where the investigated sample is water. The level of depth penetration of the IR radiation in water is marked.

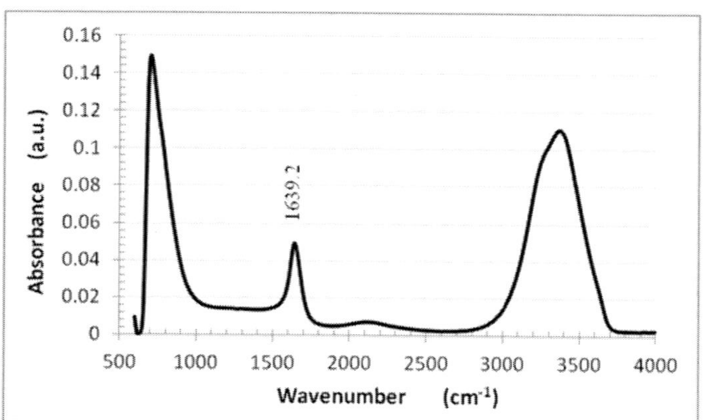

Figure 4. The ATR spectra measured for water with the specific peak absorption centered on 1639.2 cm^{-1}.

When the sample space is filled with paper, the IR-ATR spectrum looks like in Figure 5: it is characterized by the peak from 1029 cm^{-1} (λ=9718.2 nm) assigned to cellulose and the peak centered on 1416 cm^{-1} (λ= 7062.1 nm) assigned to $CaCO_3$. The absorption band specific to the OH vibrations is present due to the cellulose.

In order to determine the refractive index of paper for this specific optical domain, the Scout simulation method [13] was used. The measured spectrum was simulated (using oscillators that define the vibration modes) and the refractive index was found as n_2=1.49 corresponding to 1029 cm^{-1} (λ=9718.2 nm), and n_2=1.48 that corresponds to 1416 cm^{-1} (λ= 7062.1 nm).

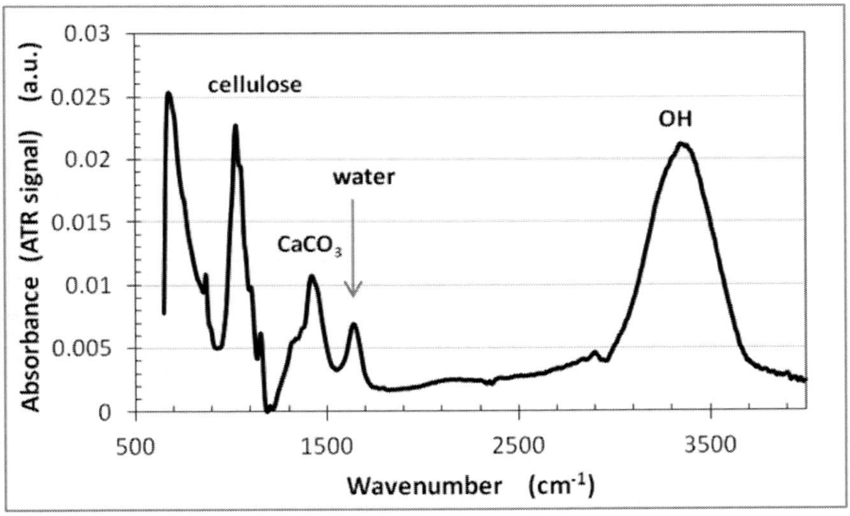

Figure 5. The IR-ATR spectrum of paper that contains $CaCO_3$ as filler.

3.2. ATR Water Probing during Penetration into Paper

If the layer of water from the experimental arrangement showed in Figure 3 is covered with porous material (e.g., paper), the water molecules will start to penetrate through the pores and leave the top of the ATR crystal. Correspondingly, the paper (let's consider it as the porous material for this work) will get closer to the ATR surface. This is equivalent with a decreasing in the water signal and an increasing in the signal assigned to paper. For more clarity, the experimental situation is presented in Figure 6 and the spectra measured at different moments of time are given in Figure 7. After 30s the signal assigned to water (1639cm^{-1}) has a smaller value when comparing with its value at t=0s; when recording the same spectrum after 2085s from the beginning of the experiment, its value is much diminished. The decreasing in the intensity of the water signal means the diminishing of the water molecule concentration in the investigated volume with the ATR technique (defined by the depth of penetration, δ_p). The presence of water in this volume is less; the water that penetrates the paper pores is replaced by the paper that gets closer to the ATR crystal surface.

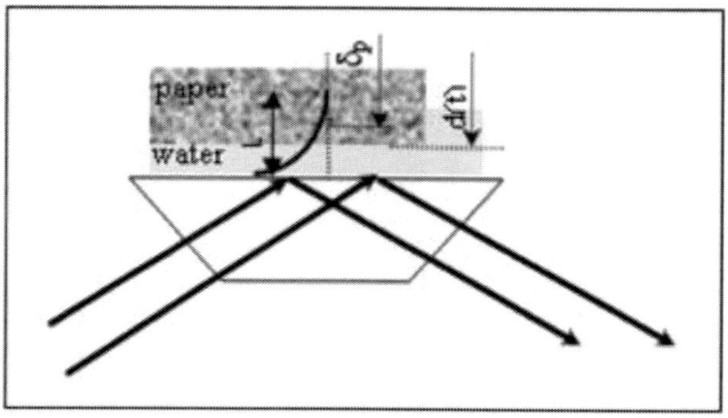

Figure 6. The experimental arrangement for studying the liquid penetration into paper. The thickness of the water layer d(t) decreases in time due to water penetration in paper pores. Correspondingly, paper gets closer to the ATR crystal surface, during the experiment.

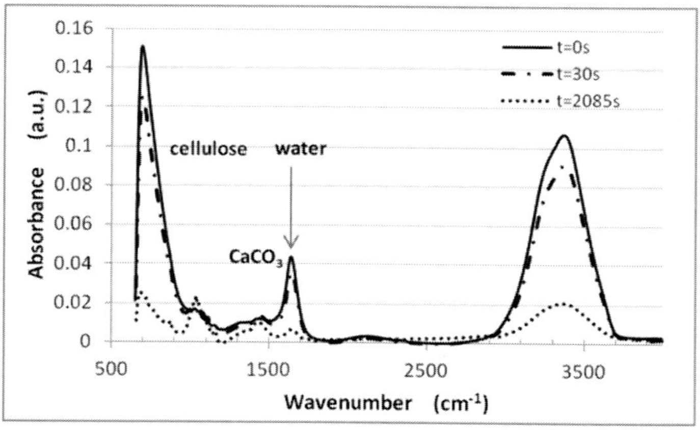

Figure 7. ATR spectra measured during the penetration of water in paper. In the legend are given three moments of time collection of the spectra. The decreasing of the ATR peak specific to water is evident.

In order to mathematically express the ATR intensity, considering the sketch from Figure 6, where the distance between the ATR crystal and the porous paper, d(t), is smaller than the depth of penetration and supposing the paper porosity $\in$, the measured ATR signal for water can be written following relation (6), as:

$$\Im(t)=\int_0^{d(t)}\chi\cdot N\cdot c\cdot\gamma\cdot E_0^2\cdot\exp\left(-\frac{2z}{\delta_p}\right)\cdot dz+\int_{d(t)}^{L}\in\cdot\chi\cdot N\cdot c\cdot\gamma\cdot E_0^2\cdot\exp\left(-\frac{2z}{\delta_p}\right)\cdot dz \tag{7}$$

where the first term represents the water in the layer thickness d(t) and the second term reveals the water in the paper pores in the (L-d(t)) thickness.

After few calculations:

$$\Im(t)=\chi\cdot N\cdot c\cdot\gamma\cdot E_0^2\cdot\left[\int_0^{L}\exp\left(-\frac{2z}{\delta_p}\right)\cdot dz-(1-\in)\cdot\int_{d(t)}^{L}\exp\left(-\frac{2z}{\delta_p}\right)\cdot dz\right]$$

and more,

$$\Im(t)=\chi\cdot N\cdot c\cdot\gamma\cdot E_0^2\cdot\frac{\delta_p}{2}\cdot\left\{\left[1-\exp\left(-\frac{2L}{\delta_p}\right)\right]-(1-\in)\cdot\left[\exp\left(-\frac{2d(t)}{\delta_p}\right)-\exp\left(-\frac{2L}{\delta_p}\right)\right]\right\}$$

or,

$$\Im(t)=\chi\cdot N\cdot c\cdot\gamma\cdot E_0^2\cdot\frac{\delta_p}{2}\cdot\left\{\left[1-\in\cdot\exp\left(-\frac{2L}{\delta_p}\right)\right]-(1-\in)\cdot\exp\left(-\frac{2d(t)}{\delta_p}\right)\right\} \tag{8}$$

This is the signal that describes the water present in the investigated volume (partially it is in the paper pores). The water that penetrates into the paper thickness (diffused and absorbed) can be revealed, at any time, from this ATR signal, when a normalized relation from Rel. (6) and Rel. (8) is written:

$$\frac{\Im_0-\Im(t)}{\Im_0}=(1-\in)\cdot\frac{\int_{d(t)}^{L}\exp\left(-\frac{2z}{\delta_p}\right)dz}{\int_0^{L}\exp\left(-\frac{2z}{\delta_p}\right)dz}=(1-\in)\cdot\frac{\exp\left(-\frac{2d(t)}{\delta_p}\right)-\exp\left(-\frac{2L}{\delta_p}\right)}{1-\exp\left(-\frac{2L}{\delta_p}\right)} \tag{9}$$

In the first approximation, $\exp\left(-\frac{2L}{\delta_p}\right) \to 0$ and, for u<1 the exponential function, exp(-u), can be written as a sum of powers of the exponent: $\exp(u) = \sum_{n=0}^{\infty} \frac{(-1)^n}{n!} \cdot u^n$, that means:

$$\exp\left(-\frac{2d(t)}{\delta_p}\right) = \sum_{n=0}^{\infty} \frac{(-1)^n}{n!} \cdot \left[\frac{2d(t)}{\delta_p}\right]^n \tag{10}$$

Considering these approximations, the ATR signal can be expressed as:

$$\frac{\Im_0 - \Im(t)}{\Im_0} = (1-\in) \cdot \left[1 - f\left(\frac{d(t)}{\delta_p}\right)\right] \tag{11}$$

with

$$f\left(\frac{d(t)}{\delta_n}\right) = \begin{cases} \frac{2d(t)}{\delta_p}, & d \ll \frac{\delta_p}{2} \\ \frac{2d(t)}{\delta_p} \cdot \left[1 - \frac{d(t)}{\delta_p} + \frac{2}{3} \cdot \left(\frac{d(t)}{\delta_p}\right)^2\right], & d \approx \frac{\delta_p}{2} \end{cases} \tag{12}$$

We mention that the liquid contained in the L-d(t) layer is diffused in the paper pores and/or absorbed into the cellulose fibers. According to relations (11) and (12), the process of diffusion/absorption can be investigated via the IR-ATR spectroscopy.

3.3. The Paper Revealed by the ATR Spectra

In this type of measurements the paper gets nearer the probed volume simultaneously with the water penetration in the paper pores. The intensity of the specific absorption peak is mathematically described in the light of the relation (6) by:

$$\Im_p = \chi_p \cdot N \cdot \int_0^{L_p} c_p \cdot \gamma_p \cdot (1-\in) \cdot E_0^2 \cdot \exp\left(-\frac{2z}{\delta_{pp}}\right) \cdot dz \qquad (13)$$

where γ_p is the radiation absorption by the paper molecular entities of concentration c_p, χ_p is the constant of the instrument for this wavelength, and δ_{pp} is the depth of penetration for the IR radiation in paper.

After simple mathematics, we get:

$$\Im_p = \chi_p \cdot N \cdot c_p \cdot \gamma_p \cdot (1-\in) \cdot E_0^2 \cdot \frac{\delta_{pp}}{2} \cdot \left[\exp\left(-\frac{2d(t)}{\delta_{pp}}\right) - \exp\left(-\frac{2L_p}{\delta_{pp}}\right)\right] \qquad (14)$$

and this means:

i) for t=0, when d(t)=L, $\Im_p = 0$;

ii) for $t=t_L$, when d(t)=0, $\Im_p = \Im_p^{max}$, where t_L is the moment when the paper is in full contact with the ATR crystal.

The $\Im_p$ can be assigned either to cellulose signal or to $CaCO_3$ signal. The maximum value of the peak intensity assigned to paper can be estimated from Rel. (14) as:

$$\Im_p^{max} = \chi_p \cdot N \cdot \gamma_p \cdot c_p \cdot (1-\in) \cdot \frac{\delta_{pp}}{2} \cdot \left[1 - \exp\left(-\frac{2L_p}{\delta_{pp}}\right)\right]. \qquad (15)$$

After some mathematics, considering $\exp(-2L_p/\delta_{pp}) \rightarrow 0$, the variation in time of the distance d(t) can be also calculated from (14) and (15) using the measured ATR signal intensity for paper:

$$d(t) = \frac{\delta_{pp}}{2} \cdot Ln\left(\frac{\Im_p^{max}}{\Im_p(t)}\right). \qquad (16)$$

4. Liquid Penetration in Porous Paper

4.1. The Paper Structure: A Brief Introduction

From the beginning we note the existence of a large diversity of paper used in printing industry. On big scale two categories are distinguished [14]: i) plain paper, and ii) coated paper. The latter, in principle, is formed in cross-section from a bulk plain paper flanked by a thin film coating of calcium carbonate like materials. The plain paper contains cellulose fibers, fillers (kaolin, clay, etc.) and color components. In the process of paper fabrication, the fibers deposited upon each other are dried to form hydrogen bonds between them. This results in a porous structure that has a distribution of pores with various diameters which will influence the paper mechanical properties. A high resolution scanning electron microscopy (HRSEM) picture made on a cross-section of plain paper reveals the fibers, the fillers and the pores (see Figure 8). Of course the distribution of these constituents has to be seen as an average, because it is a function of the place where the cross-section is made. Such HRSEM pictures are used to show the paper complexity as a porous material. With special care, information about the porosity can be obtained. But, measurements methods as picnometry with Hg or He are used to determine the porosity parameters of paper. In the plain paper, the liquid penetration determines the growing in the cellulose fiber dimension due to its hygro- and hydroscopic properties and part of the hydrogen bonds between fibers are broken. This means that the paper structure changes during the liquid penetration.

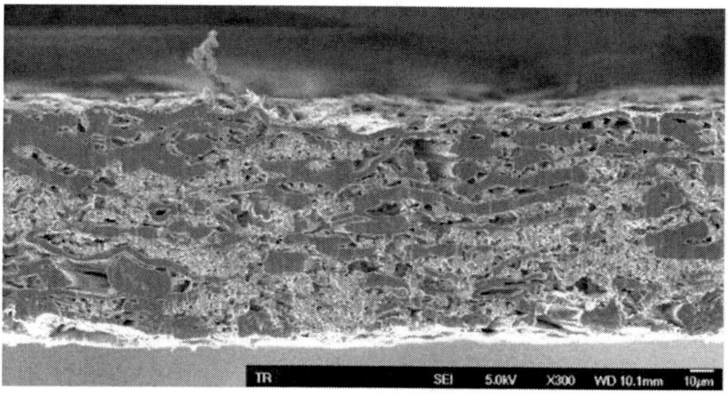

Figure 5. A HRSEM spectrum of plain paper that reveals its internal structure: the fibers (dark grey), the fillers (light gray) and the pores (complete black spots).

The porosity, the distribution, the shape and the radius of the pores determine the penetration of liquid in paper. The pores can be connected between them forming capillaries of different shapes and lengths. The surface energy of the fibers corroborated with the surface tension of the liquid defines the contact angle of the liquid/fiber interface. Treating chemically the fibers' surface to make them more hydrophilic or hydrophobic has a great influence on the driving forces of the liquid penetration. The same idea is valid for the coating material where the pores are smaller in diameter and could be more uniform. The thickness of the coating layer is non-uniform because of the roughness of the plain paper that forms the bulk.

When the studied paper is coated with a compound (kaolinite, calcium carbonate, talk, resins, etc.) to improve certain qualities, the liquid transport through it is also influenced. The porosity of the coating is different from that of the plain paper (or bulk of the coated paper). The pores' diameter is between 1-10μm, while in the coating is in the order of magnitude of 100nm. We note that in the coating, the solid skeleton is considered non-deformable during the wet status, while the fibers from the plain paper are characterized by nano-pores that uptake liquid.

The main message of this section is that the paper structure is complex and the permeation of the liquid into this porous material has to be seen like a tensor, being dependent on the investigated direction: in the paper plan (also here we distinguish between the machining direction in the fabrication process and perpendicular to it) or transversal to the paper plan. The latter is interesting for the printing industry. However, for understanding the paper properties, the anisotropy and inhomogeneity of the paper structure have to be considered.

4.2. On the Liquid Penetration in Porous Media

When a liquid gets in contact with a porous substrate it could experience: diffusion through the pores and absorption into the pores' walls (fibers if the substrate is paper or textile). These processes are more complicated considering the fact that the liquid absorption into the fibers changes the structure of the pores and the mechanical properties of the substrate. For the beginning, the absorption into the pores' walls is not considered by such kind of effects. In other words the skeleton of the solid material that defines the pores will remain unchanged during the liquid penetration through the pores. Between fibers are formed capillaries.

The liquid molecules nearby the capillary walls interact with the molecules of the same liquid – we call these interactions as cohesive, and with molecules of the fibers (solid surface) – we call these interactions as adhesive. The energy balance of the interactions at the interface between one liquid and one solid (substrate) generates the contact angle, θ. Considering the capillary situation with liquid in the capillary volume, when the cohesive forces are smaller than the adhesive forces, the contact angle is smaller than 90°and the front of the liquid/gas interface advances. The liquid penetration into these capillaries is of importance in this section. In literature there are models that describe the penetration s of a liquid with viscosity η, mass-density ρ, surface tension σ and contact angle θ in a porous material of porosity ∈ with pore radius r. The classical approach was proposed by Bosanquet [15] where the capillary, viscous and inertial forces, are considered in the framework of the well known Newtonian mechanics. The resulted equation is given in Rel. (17) and, we will consider the solution proposed by Schoelkopf et coworkers [16]:

$$2\pi \cdot r \cdot \sigma \cdot \cos\theta + P_e \cdot \pi r^2 - 8\pi \cdot \eta \cdot s\frac{ds}{dt} = \frac{d}{dt}\left(\pi r^2 \cdot \rho \cdot s\frac{ds}{dt}\right) \tag{17}$$

or,

$$\frac{d}{dt}\left(s\frac{ds}{dt}\right) + p \cdot s\frac{ds}{dt} - q = 0$$

where

$$p = \frac{8\eta}{r^2 \cdot \rho}; \quad q = \frac{P_e r + 2\sigma \cdot \cos\theta}{r \cdot \rho}. \tag{18}$$

In order to have all parameters defined in the equation (17), we note that P_e is an external pressure that can be applied to influence the liquid movement in the considered capillaries. The solution of the second degree differential equation is:

$$s^2(t) - s^2_{t=0} = \frac{2q}{p}\left[t - \frac{1}{p}\left(1 - e^{-p \cdot t}\right)\right] \quad (19)$$

and it shows a complicated time dependency of the capillary length with liquid, s.

A better image of this solution we get when $p \cdot t << 1$ (which means small viscosity (η) and large mass density (ρ) at the beginning of the liquid movement - small time (t)) and the exponential can be developed in series:

$$e^{-x} = \sum_{n=0} \frac{(-x)^n}{n!} \cong 1 - x + \frac{x^2}{2}$$

After few mathematics and approximations (i.e., $P_e=0$) the solution is written as:

$$s = \sqrt{\frac{2 \cdot \sigma \cdot \cos\theta}{r \cdot \rho}} \cdot t \quad (20)$$

and it describes the so-called inertial flow. The length of the liquid within the capillary is proportional to the time as parameter of this transport. This means a constant speed in time of the advancing front of the liquid, which is larger for the liquids with high surface tension and low mass density. The speed is inverse proportional to the capillary radius; it does not depend on the liquid viscosity. Therefore in this regime, the liquid penetrates more rapid in narrower capillaries.

At the other extreme, $p \cdot t >> 1$, after some calculations and approximations (i.e., $P_e=0$) the solution is immediately:

$$s = \sqrt{\frac{\sigma \cdot \cos\theta}{2 \cdot \eta} \cdot r} \cdot t^{1/2} \quad (21)$$

which represents the well known Lucas-Washburn relation [17], [18].

This relation shows a capillary length filled with liquid proportional to the time at the power 0.5, which means that the liquid advancing speed is time dependent. Moreover, the s parameter (capillary length with liquid) is larger for liquids with high surface tension and low viscosity; it is not dependent on the liquid density. Interesting to note is that the front liquid advances faster in capillaries with larger radius.

The general solution (see the relation (19)) of the equation and its particular expressions given by relations (20) and (21) are important to understand the time dependency of the capillary rise, and how the liquid properties influence it. Schoelkopf et al. pay special attention to this aspect especially in the figure 1 of their paper [16]. The authors illustrated quantitatively the length "s" for 1,3-propandiol in tubes of 1 μm and 1mm radius. The inertia is larger for liquids with larger mass density entering a large radius capillary. With the courtesy of J. of Colloids and Interface Science we reproduce their results in the Figure 9 with the following outputs: a) the flow of the liquid changes from the inertial-like (relation (20)) to the Lucas-Washburn -like flow (relation (21)) as "t" and "s" increase; b) for a capillary with 1 μm radius, the change happens at t=10^{-8}s (experimentally it is difficult to measure it); c) for tubes with r=1mm the changeover occur at 10^{-2}s (this is an accessible time for measurements).

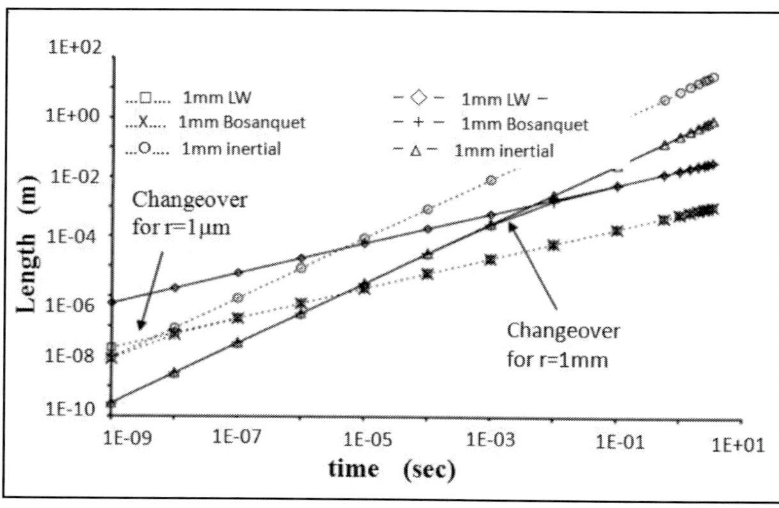

Figure 9. The time scale for the two regimes: inertial flow and Lucas-Washburn equation calculated for two capillary radius values: r=1μm and r=1mm. This plot is reproduced with the courtesy of J. of Colloids and Interface Science.

According to the same study, the large capillaries are dominated by the inertial retardation effects, while small capillaries are faster filled with liquid.

Concerning the speed of the advancing front of liquid, this parameter can be calculated using relations (20) and (21) and applying the classical definition:

$$v = \frac{ds}{dt} = \begin{cases} \sqrt{\dfrac{2 \cdot \sigma \cdot \cos\theta}{r \cdot \rho}} & \text{inertial flow} \\ \dfrac{\sigma \cdot \cos\theta}{4\eta} \cdot \dfrac{r}{s} & \text{Lucas - Washburn flow} \end{cases} \tag{22}$$

To conclude this theoretical part of our subject, according to [16] we can say: for capillaries with radius $r \in (0.1\mu, 100\mu)$ the Lucas-Washburn equation is the one which describes the phenomenon of liquid penetration considering the 0.01 sec measurement time scale. A detailed analysis of the classical theory of capillarity showing different limiting cases of one rather general equation (similar to (17)) is given in “dynamics of capillary rise” of Zhmud et.al. [19]

Most of the publications that treat the liquid propagation in porous paper, based on both experimental results and computational modeling, reveal the square root dependency on time of the liquid path in the pores. In this light, hereunder some models are summarized:

i. Darcy[‡]: $s = \sqrt{\dfrac{4K \cdot \sigma \cdot \cos\theta}{\eta \cdot r}} \cdot t^{1/2}$, with K = Darcy’s constant

ii. Carman-Kozeny [20], [21] where the Darcy’s coefficient is described by the porosity and a parameter k_1: $K = \dfrac{\in^3}{k_1 \cdot (1-\in)^2}$.

iii. Schuchardt-Berg [22]: $s = \sqrt{\dfrac{\sigma \cdot \cos\theta}{2\eta} r_0} \cdot \left[t - \dfrac{a}{r_0} t^2 + \dfrac{a^2}{3r_0^2} t^3 \right]^{1/2}$, where the parameter “a” describes the decreasing of the pore radius due to swelling: $r = r_0 - a \cdot t$;

iv. Massodi – Pillai [23]: $\ln\left(\dfrac{s_0}{s}\right) = \dfrac{K\rho\rho g_w^2}{\eta L_w R_s^2} \cdot t$, where r_w and L_w are the wick radius and length, respectively;

v. Hassanizadeh – Gray [24] where the movement of the meniscus and the contact line formed in a capillary tube is described considering the

‡ Darcy's equation is phenomenologically derived and describes the flow of a fluid through a porous medium. Henry Darcy has formulated it in 1856 based on the results of his experiments on the water flows through sand beds. This equation is the scientific basis of fluid permeability used in the earth sciences, particularly in hydrogeology.

effect of the surface energy of the meniscus into the thermodynamically description of capillary pressure at micro-scale level.

One issue is clear from all these models, and others not listed here: the time dependence of the liquid advancing into a capillary, according to the publications, is either $\sim t^1$ (inertial flow) or $\sim t^{1/2}$ (capillary forces). The coefficients that multiply these time functions are defined by the liquid properties, by the paper characteristics and by their interactions. In most of the models, these time dependencies have been deduced theoretically considering the matrix of the pores (skeleton) as being non-deformable solid. That is not the case for paper. In order to be able to study the liquid penetration in porous paper materials, a very general function is here proposed:

$$s = \Psi(\text{liq}, \text{pap}, \text{int}) \cdot t^{\alpha} \qquad (23)$$

where the parameter Ψ is formed by all properties that are time-independent, while α describes the total dependency on time. The latter parameter is, in our view, describing the mechanism of liquid penetration.

4.3. From the IR-ATR Signal to the Liquid Penetration Length

Supposing that the liquid penetrates vertically in the porous material, let's consider that the paper bottom moves towards the ATR crystal on the distance L-d(t) (see the Figure 6). In this situation, the liquid volume considering the paper porosity can be expressed as:

$$A \cdot [L - d(t)] \cdot (1 - \in) = s(t) \cdot \pi \cdot r_{eq}^2$$

where r_{eq} is an equivalent radius of the pores. The parameter of interest is d(t):

$$d(t) = L - s(t) \cdot \frac{\pi \cdot r_{eq}^2}{A \cdot (1 - \in)} = L - \Psi \cdot \frac{\pi \cdot r_{eq}^2}{A \cdot (1 - \in)} \cdot t^{\alpha} \qquad (24)$$

Introducing (24) in Rel. (11) and (12) we get:

$$\frac{\Im_0 - \Im(t)}{\Im_0} = (1-\in)\cdot\left[1-\frac{2}{\delta_p}\cdot\left(L-\Psi\cdot\frac{\pi\cdot r_{eq}^2}{A\cdot(1-\in)}\cdot t^{\alpha}\right)\right] = \\ = (1-\in)\cdot\left[1-\frac{2L}{\delta_p}+\frac{2}{\delta_p}\cdot\Psi\cdot\frac{\pi\cdot r_{eq}^2}{A\cdot(1-\in)}\cdot t^{\alpha}\right] \tag{25}$$

Therefore, in the first approximation the signal $\frac{\Im_0-\Im(t)}{\Im_0}$ is proportional with the penetration length into an equivalent pore modeled by $\Psi\cdot t^{\alpha}$.

5. VALIDATION OF THE MODEL WITH EXPERIMENTAL DATA

In order to verify the ATR model presented in the previous sections, a confrontation with the experimental data is needed. The ATR spectra measured during the water penetration through the pores' of a plain paper are shown in the Figure 7. As it was already specified, in this figure there are shown spectra measured during the water penetration in paper. In this section we will focus on the absorption peaks that reveal information about water molecules -1639 cm^{-1} and the cellulose - the large band that peaks [25] at 1059 cm^{-1}. The peak height values for these two signals during the experiments are shown in Figure 10.

The evolution in time of the water signal is characterized by: a) a small period (up to 15s) when the signal decreases very slowly (it is quasi constant) in time; b) a rapid decreasing region, and c) a slow decreasing region that ends asymptotically constant. Regarding the cellulose signal, the trends are: i) a rapid increasing in time, ii) followed by a slow asymptotically increase towards a constant maximum value.

A closer look on the data of the water signal reveals that its variation in time is more complicated. The insert of Figure 10, where the experimental raw data are plotted in log-log scales, emphasizes the differences in the time evolutions of the recorded ATR signals for water and cellulose. Before any discussion, we mention that the penetration depth of the specific IR wavelength in cellulose is with 65% larger than the depth of penetration of the specific wavelength in water ($\delta_{pp}/\delta_{pw} = 1.65$). This means that for a while, the signal assigned to cellulose will monotonically increase and the water signal will be constant. Indeed, the representation in the log-log scale shows for t<15s a cvasi-constant value of the ATR signal corresponding to $1639 cm^{-1}$. According to the theoretical description of the ATR measurement, this situation occurs when the thickness of the water layer on the ATR crystal is larger than $L \sim 3\delta_p$. Of course in this interval of time the water molecules continued their penetration in paper pores. Also, the time of 15s depends on the initial thickness of the water layer on the ATR crystal and on the experimental conditions.

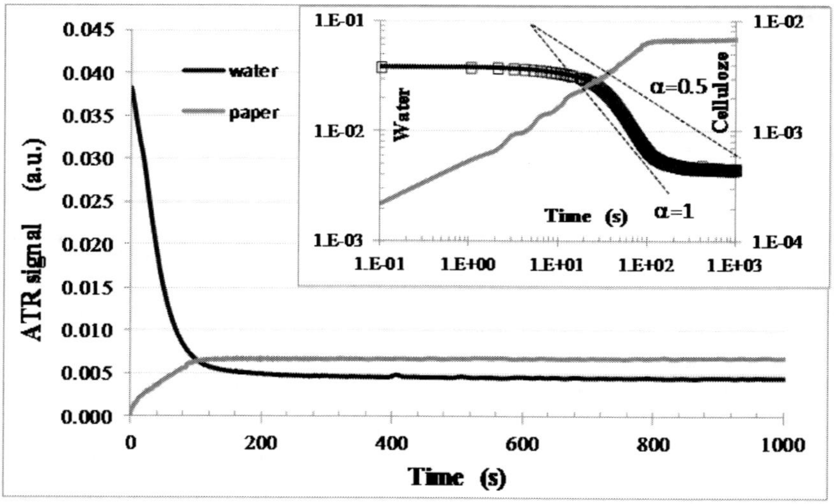

Figure 10. The IR-ATR peak intensity for water molecules ($1650 cm^{-1}$) - black line and cellulose ($1058 cm^{-1}$) in grey line, as a function of time, during water penetration in plain paper. In the insert, the same experimental data are presented in log-log scales.

For this experiment, after t=15s the water ATR signal begins to decrease with a time dependent rate. To guide the eye, in the insert of Figure 10, are shown the slopes for α=-0.5 and α=-1 in the region where the signal decreases. There are intervals of measurement time where partially we can consider these values.

In order to determine the α value, let's consider relation (25) which models the penetration of liquid in porous paper using representation $(\Im_0 - \Im(t))/\Im_0 \propto \Psi \cdot t^{\alpha}$. We note that $(\Im_0 - \Im(t))/\Im_0$ represents the normalized signal assigned to the liquid that leaves the probed area by the IR-ATR radiation. In other words, this signal shows indirectly the liquid penetration in porous paper during the experiment. Such a plot is shown in Figure 11 for the data presented in Figure 10. The fit with a power function in the log-log scales produces the α value as 1.07, which is valid for a long measurement time.

In order to have a measure of the other parameter from relation (25) – the pre-factor Ψ – we propose to represent the $(\Im_0 - \Im(t))/\Im_0$ as a function of t^α. The slope of the linear part characterizes the parameter Ψ. According to Figure 12 this slope is 9.44 10^{-3} $sec^{-\alpha}$. Sure, from relation (25) the slope is $\frac{2}{\delta_p} \cdot \frac{\pi \cdot r_{eq}^2}{A} \cdot \Psi$ and the equivalent radius of the pores is difficult to be estimated. This depends on the porosity, on the number of pores and the averaged diameter of the pore. But, this slope value is useful when the same paper is used in measurements made under different conditions; in this way, the variation is due to the Ψ parameter. Also, more about the influence of the liquid nature on the penetration mechanism can be obtained studying with this method the penetration of different liquids in the same paper type. These data are given in the second part of this study [9].

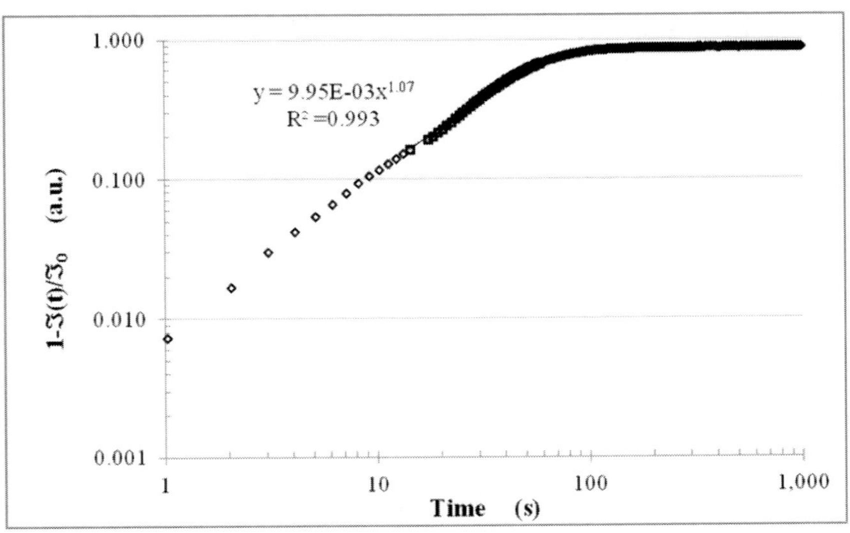

Figure 11. The signal $(\Im_0 - \Im(t))/\Im_0$ which reveals the amount of liquid that leaves the space probed by the ATR, as a function of time.

Relevant information that can be obtained from this model is related to the penetration rate of the liquid in paper. According to relation (22) this rate can be calculated from the time dependency of the length of the liquid in paper. Supposing that the front of advancing fluid is the same on the whole area, from relation (24) we can say that this rate is measured by speed of paper movement towards the ATR crystal:

$$|v| = \left|\frac{d}{dt}d(t)\right| = \frac{\pi \cdot r_{eq}^2}{A(1-\in)} \cdot \frac{d}{dt}s(t) = \zeta \cdot \frac{\delta_p}{2(1-\in)} \cdot \alpha \cdot t^{\alpha-1} \quad (26)$$

where $\zeta = \frac{\pi \cdot r_{eq}^2}{A} \cdot \Psi$ is the slope of the $(\Im_0 - \Im(t))/\Im_0$ versus t^α plot.

For situations when α=1, this rate of penetration is constant in time; in all other situations, it is time dependent (see relation (22)).

The water penetration rate obtained from the data here analyzed (Figure 10) using this model amounts to 4.7 10^{-3} units/s. [§]

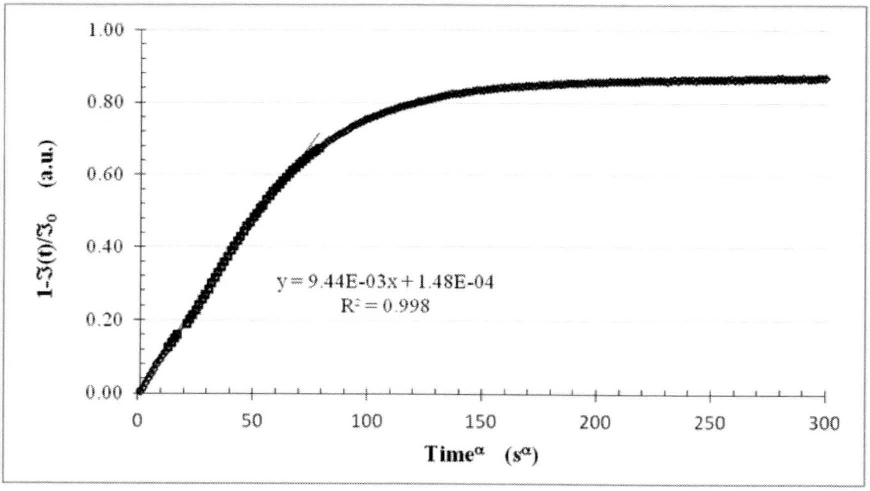

Figure 12. The Ψ value is obtained from the slope of the $(\Im_0 - \Im(t))/\Im_0$ versus t^α plot.

What about the paper signal? How this fits into this model? Answering these questions means to compare the time dependencies of the water thickness d(t) on the ATR crystal obtained using the water signal (relations (11) and (12)) and the paper signal (relation (16)), respectively. The results of

[§] The rate of propagation has to be expressed in units of length, or mass per second. In order to have this, a calibration of the IR measurements it is necessary in advance, to correlate the IR intensity to the amount of liquid. This calibration has to be made for each investigated liquid.

the calculations for the normalized thickness $d(t)/\delta_p$ are shown in Figure 13. We note that: a) the normalization is made to the penetration depth specific for each assigned signal (water and paper, respectively); b) the initial moment for water signal is considered for t=15sec, when the signal decreases (the penetration in paper pores becomes measurable with this signal). We consider the results obtained from analyzing the two different signals as being in good agreement, which strongly supports the proposed model to study the liquid penetration in porous materials using the IR-ATR spectroscopy. The difference that appears for 16s<t<50s could be due to the paper skeleton deformation induced by the water propagation, mechanism that is not considered in the model. This is a supposition that has to be validated.

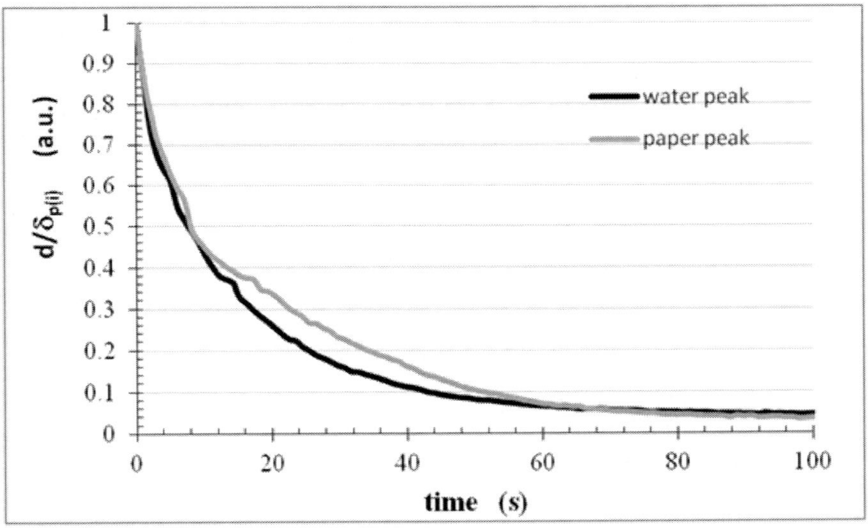

Figure 6. The thickness of the water layer above the ATR crystal determined from the IR-ATR signals assigned to water and paper, measured simultaneously.

The porosity of the paper is included in relation (16); for the samples of plain paper used in this study the porosity is 0.57. The influences of the liquid properties and of the experimental conditions (temperature and relative humidity) on the penetration mechanisms are revealed in part II [9] published herewith.

CONCLUSION

Studying the liquid penetration in a porous material with a "solid skeleton" that changes during this process is challenging. In this work the attenuated total reflection (ATR) spectroscopy in infrared optical range is proposed as a method to investigate the liquid transport through the porous paper. The ATR method is presented with many details; a brief introduction of the models most used in literature to describe the liquid penetration in porous materials is also given. A new approach of the IR-ATR data modeling is proposed with a parameterization that, in our view, is general and, at the same time, reveals the main features of the liquid penetration. From the time evolution of the IR signals assigned to water or paper, the liquid penetration in porous paper is modeled as $\Psi \cdot t^{\alpha}$. The total time dependency of the measured signal is described by the α parameter while the factor Ψ includes the influences of liquid properties, paper characteristics and their interactions. Considering the model here presented together with the analysis of the experimental data (see the part II of the study) we conclude that the IR-ATR is a tool to investigate the liquid penetration in porous materials under various circumstances (temperature, relative humidity, porosity levels, etc.).

REFERENCES

[1] M. Alava and K. Niskanen, "The physics of paper," *Rep. Prog. Phys. 69,* pp. 669-723, 2006.

[2] J. Verhoeff, J. Hart and W. Gallay, "Sizing and the mechanism of penetration of water into paper," *Pulp and Paper Magazine of Canada, 64 (12),* pp. T509-516, 1963.

[3] J. A. Bristow and H. Bergenblad, "Interpretations of ink-stain tests on coated papers.," *Adv. In Printing Sci. and Tech. 16* , pp. 373-391, 1982.

[4] T. Helle and P. Q. Johnsen, "Using stereoscopic and SEM backscatter imaging for studying ink distribution details on paper and fibre surfaces," *J. Pulp Pap.Sci. 20 (7),* pp. 189-194, 1994.

[5] M. Daun, "Model for the dynamics of liquid penetration into porous structures and its detection with the help of changes in ultrasonic attenuation.," Darmstadt University, Darmstadt, 2006.

[6] C. M. Tag, M. Juuti, K. Koivunen and P. A. Gane, "Dynamic water transport in pigmented porous coating medium: novel study of droplet

absorption and evaporation by near-infrared spectroscopy.," *Ind. Eng. Chem. Res.* , vol. 49, pp. 4181-4189, 2010.

[7] T. -Q. Li, M. Haggkvist and L. Odberg, "The porous structure of paper coatings studied by water diffusion measurements.," *Colloids and Surfaces A: Physicochemical end Engineering Aspects 159,* vol. 159, no. 1, pp. 57-63, 1999.

[8] T. Lamminmaki, H. Lampinen, S. Hyvarinen and K. Kataja, "Measuring the penetration properties of starch-trated wood-free paper.," *Nordic Pulp and Paper Research Journal, 25,* pp. 372-379, 2010.

[9] N. Tomozeiu and H. Boonen, "IR attenuated total reflection – a tool to investigate liquid penetration in paper: experimental.," in *Optical Spectroscopy: Technology, Properties and Performances*, Nova Publisher, 2014, p. ch. 5.

[10] J. Hecht, Optics: light for a new age, Cornell University: Scribner, ISBN: 0684188791, 9780684188799, 1988.

[11] M. Katz, Introduction to geometrical optics, World Scientific, ISBN 9812382240, 9789812382245, 2002.

[12] F. M. Mirabella, "Strength of interaction and penetration of infrared radiation for polymer films in internal reflection spectroscopy.," *Journal of Polymer Science: Polymer Physics* , vol. 21, no. 11, pp. 2403-2417, 1983.

[13] W. Theiss, "www.wtheiss.com," Hard- and Software, 2010. [Online].

[14] L. Pal, M. K. Joyce and P. D. Fleming, "A simple method for calculation of the permeability coefficient of porous media.," *TAPPI Journal, Vol. 5 No 9,* pp. 10-16, 2006.

[15] C. M. Bosanquet, "On the flow of liquids into capillary tubes.," *Philosophical Magazine, Series 6, 45, Issue 267,* pp. 525-531, 1923.

[16] J. Schoelkopf, C. J. Ridgway, P. A. Gane, G. P. Matthews and D. C. Spielmann, "Measurement and network Modeling of liquid permeation into compacted mineral blocks.," *Journal of Colloid and Interface Science 227,* pp. 119-131, 2000.

[17] R. Lucas, "Rate of capillary ascension of liquids.," *Kolloid Z. 23,* pp. 15-22, 1918.

[18] E. W. Washburn, "The dynamics of capillary flow.," *Phys. Rev. Vol 17, Issue 3,* pp. 273-283, 1921.

[19] B. V. Zhmud, F. Tiberg and K. Hallstensson, "Dynamics of capillary rise.," *Journal of Colloid and Interface Science, 228,* pp. 263-269, 2000.

[20] P. C. Carman, "The determination of the specific surface of powders.," *J. Soc. Chem. Ind., 57,* pp. 225-234, 1938.

[21] J. Kozeny, "Uber kapillare Leitung des Wassers in Bodem.," *Royal Academy of Science, Vienna, Proc. Class I, 136,* pp. 271-306, 1927.

[22] D. R. Schuchardt and J. C. Berg, "Liquid transport in composite cellulose superabsorbent fiber network.," *Wood Fiber Science 23,* pp. 342-357, 1990.

[23] R. Masoodi, K. M. Pillai and P. P. Varanasi, "Darcy's law-based models for liquid absorption in polymer wicks.," *Am. Inst. Chem. Eng. Journal, 53,* pp. 2769-2782, 2007.

[24] S. M. Hassanizadeh and W. G. Gray, "Thermodynamic basis of capillary pressure in porous media.," *Water Resources Research, Vol. 29, No. 10,* pp. 3389-3405, 1993.

[25] D. Ciolacu, F. Ciolacu and V. I. Popa, "Amorphous cellulose – structure and characterization.," *Cellulose Chemistry and Technology, 45 (1-2),* pp. 13-21, 2011.

In: Optical Spectroscopy
Editor: Nicolae Tomozeiu
ISBN: 978-1-63321-197-1

Chapter 5

IR ATTENUATED TOTAL REFLECTION: A TOOL TO INVESTIGATE LIQUID PENETRATION IN PAPER: EXPERIMENTAL

Nicolae Tomozeiu [*] ***and Hennie Boonen***
A&PS, R&D Océ Technologies, Venlo, The Netherlands

ABSTRACT

Using the Attenuated Total Reflection method of the infrared spectroscopy (IR-ATR), the water penetration from thin liquid layers in porous paper is investigated. The paper structure is a porous system consisting of cellulose fibers that interact via hydrogen bonds, fillers and voids (pores). Two methods based on IR-ATR spectroscopy are designed and used in this work: (i) the liquid is placed on the ATR crystal while the paper covers it, and (ii) the paper lies on the ATR crystal and the liquid is placed on top.

A new model proposed to describe the liquid penetration is applied for analyzing the experimental data obtained under various environmental conditions (relative humidity and temperature). The influences of the paper thickness and of the liquid nature on the liquid penetration rate are also revealed.

Keywords: Liquid penetration in paper, IR-ATR spectroscopy

---

[*] Corresponding author address: Nicolae.Tomozeiu@Oce.com.

1. INTRODUCTION

The drying of the water based inks printed on porous substrates is dependent on the evaporation phenomena and the liquids' penetration in the media. Several parameters like temperature, relative humidity of the environment, viscosity and polarity of the liquid as well as porosity and type of the paper influence the diffusion / absorption processes of the liquid penetration. To better understand these processes, two infrared spectroscopy methods based on attenuated total reflection (ATR) spectroscopy are proposed in this work. The theoretical background of the hereunder discussed experiments is given in the twin-chapter titled "IR attenuated total reflection – a tool to investigate liquid penetration in paper: theoretical considerations" [1]. We note that there are many publications concerning the liquid penetration in/through porous media [2-4]. Both phenomenological and theoretically deduced formulas that express the liquid penetration rate have been presented and argued in this literature, under various conditions (e.g., the vertical capillaries of the pores, the constant surface tension force that acts in the pores, etc…). The paper, as a porous media, is a difficult material to study the liquid penetration into it due to the non-uniform distribution of the physical/chemical properties and their anisotropy. As a result of its fabrication processes, the paper reveals a large difference concerning its properties (mass density, fibers distribution, pores dimension and spatial shape, etc.) both in the plane of the paper and perpendicular on this plane (in its thickness). Due to the ingredients utilized to obtain the desired properties for paper, the interaction with the liquid is strongly influenced by the processes at micro- scale and their integration into the measurable parameters, which are defined at a macro-scale level. Moreover, we stress that during the liquid interaction with the pores' walls (the cellulose fibers) the pores shape changes and the internal structure of the paper modifies due to hydro- and hygroscopic properties. Therefore the liquid penetration in/through porous paper is a dynamic process.

However, in many industrial areas, understanding the interaction processes between a liquid and paper is very important. In most of the applications, the quality and the properties of the final product depend on the rate of the liquid penetration in pores, on the interaction at atomic/molecular level between the liquid and the fibers that design the porous structure. Measuring the amount of liquid that penetrates the paper, considering the environmental conditions (e.g., temperature, relative humidity) and understanding the physical/chemical processes that determine the penetration rate are important and non-trivial tasks.

In this work we will analyze experimental data obtained from infrared attenuated total reflection (IR-ATR) methods and presented in the light of the model from the previous chapter [1].

2. EXPERIMENTAL METHODS

All spectra have been measured on a Bruker Vertex 80 spectrometer with a PIKE veeMAX-II accessory and a heatable Ge-crystal with angle θ=45°. The temperature of the crystal is controlled with a PIKE temperature controller. If not specially mentioned, the temperature of the ATR crystal during the measurement was 25 °C.

As it was explained in the previous chapter, the sample holder in an IR-ATR experiment is the ATR crystal. The sample is probed by the IR beam that via evanescent waves penetrates a well defined depth into the material. The measure of this depth of penetration is given by relation (4) of [1] and it depends on the molecules' type (via the wavelength) and the material optical properties (refractive index). The intensity of the measured signal can be calculated with relation (6) of the previous chapter. Based on these, we have designed two methods to investigate the liquid penetration in/through the porous paper:

i) The liquid is layered on the ATR crystal and it is covered with the porous paper; the liquid signal is maxim at the beginning and it decreases in time because the liquid gets out from the probed volume going into the porous paper.
ii) The paper is put on the ATR crystal and the liquid on top of it; the liquid signal appears and increases in time when the liquid gets closer and closer to the ATR crystal through the pores and cellulose fibers.

Schematically the two methods are presented in Figure 1. As we have shown in the previous chapter [1], in the case of method (i) the IR-ATR signal specific to the liquid molecules measures the liquid disappearance from the probed volume (the information about liquid penetration in pores is indirectly obtained). When the liquid is applied on the paper that lies on the ATR crystal - method (ii) - the ATR signal will reveal its appearance in the investigated volume.

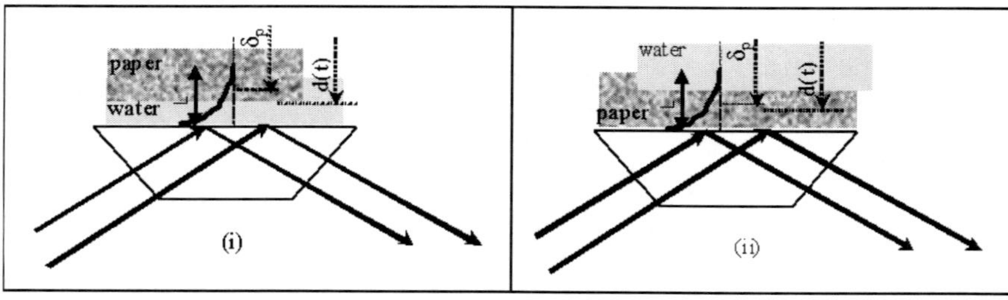

Figure 1. Two methods based on IR-ATR spectroscopy that reveal the liquid diffusion and absorption through/into porous paper.

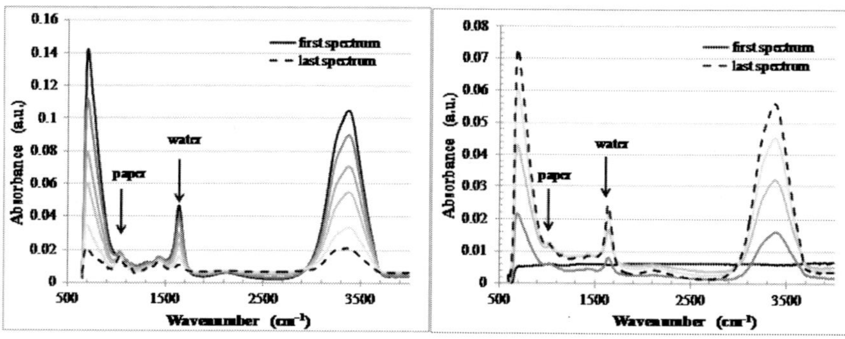

Figure 2. ATR spectra measured with method (i): the peaks related to water decrease in time while the peak assigned to paper increases and, with method (ii): the time evolution of the peaks intensity is reverse to method (i). The black continue spectrum is the first measured and the black dashed spectrum is the last one measured.

We consider instructive to show spectra collected via these two methods at different moments of time. As it can be seen in Figure 2 the signal assigned to water molecules (1639 cm^{-1}) decreases in time in the case of the experiments with method (i), while the same signal measured with method (ii) increases in time. This means that the presence of water in the probed volume diminishes in time and increases, respectively. In the same way the peak assigned to paper (1029cm^{-1}) that increases in the first experiment and remains quasi-constant in the second experiment can be analyzed.

3. A Comparison between the Two Methods

For comparison let's consider the water ATR signal measured by the height of the peak absorption centered on 1639 cm^{-1}. The data are shown in

Figure 3, where with black color is plotted the evolution in time of the water signal measured with method (i) and with grey color is shown the water signal measured via method (ii) versus time.

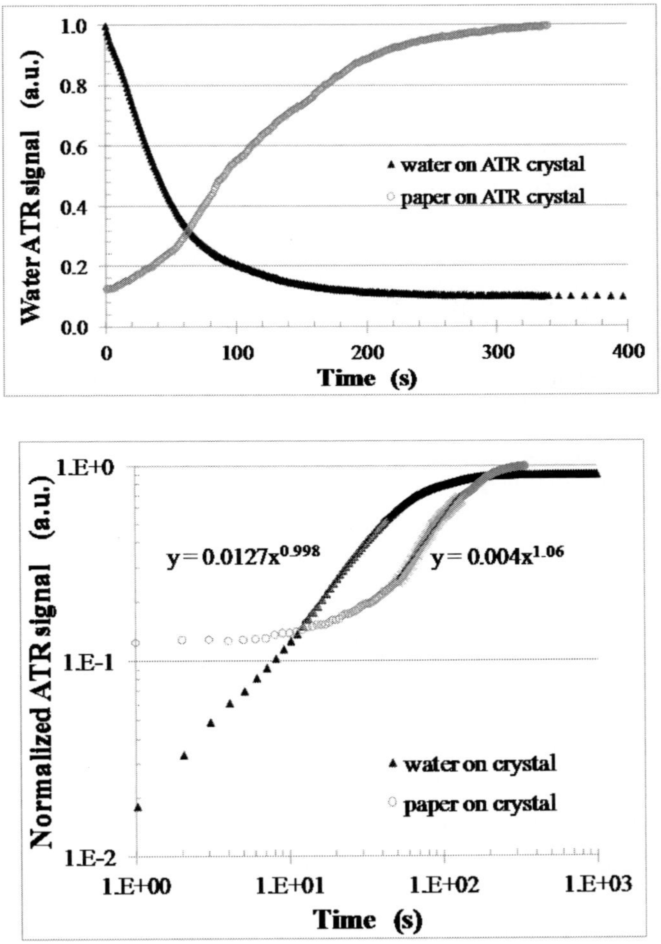

Figure 3. The normalized peak height of the water ATR signal versus time: a comparison of the two methods of investigation. In the lin-lin scale plot are shown the evolutions in time of the measured ATR signal assigned to water. In the log-log scale plot, according to the models proposed in this work, the penetration of water in porous paper is revealed (see the explanations in text).

As it was already mentioned, one plot shows the disappearing of water from the ATR probed volume and the other one reveals the water appearing in this volume. In the first case (measurement method (i)), the water molecules disappear due to their propagation in paper placed on top of the water layer (see Figure 1). These data can be used to indirectly study the liquid penetration in porous paper. The model utilized to calculate the water penetration in paper using the ATR signal is represented by relation (24) in [1]:

$$\frac{\Im_0 - \Im(t)}{\Im_0} = (1-\in)\cdot\left(1-\frac{2L}{\delta_p}\right)+\frac{2}{\delta_p}\cdot\Psi\cdot\frac{\pi\cdot r_{eq}^2}{A}\cdot t^{\alpha} \tag{1}$$

where $\Im_0$ and $\Im(t)$ are the signal intensities when the whole probed volume is water and its value in time during the penetration in paper, respectively; δ_p is the depth of penetration of the IR radiation into water layer, L ~$3\delta_p$ is the experimentally proved depth [5] that contributes to the ATR signal, $\in$ is the paper porosity, A is the area of the IR spot used in the ATR measurements, r_{eq} is an equivalent radius of the pores, Ψ and α are parameters that describe the liquid transport in porous paper. It is immediately seen that the ATR signal is normalized to its initial value.

The experimental data of method (ii) show "live" the water penetration in paper and in the light of the model presented in [1] they can be mathematically expressed as:

$$\begin{aligned}\Im(t) &= \chi\cdot N\cdot c\cdot \in\cdot\gamma\cdot E_0^2\cdot\frac{\delta_p}{2}\cdot\left\{\exp\left[-\frac{2d(t)}{\delta_p}\right]-\exp\left(-\frac{2L}{\delta_p}\right)\right\}\cong \\ &\cong \chi\cdot N\cdot c\cdot\gamma\cdot\in\cdot E_0^2\cdot[L-d(t)] \quad \text{for } d(t)\le L\end{aligned} \tag{2}$$

where χ is a measurement-setup constant, γ is the absorption coefficient specific to the radiation wavelength, c is the concentration of the investigated species and d(t) is the non-wet thickness of the paper placed on the ATR crystal. For $d(t) \geq L$ the recorded ATR signal is zero and it increases in time once d(t) becomes smaller than L~$3\delta_p$. Normalizing the data to the moment when d(t)=L, t=0, it is clear that $\Im(t)$ increases in time. Considering a hypothetical capillarity of an equivalent radius r_{eq}, we can write:

$$[L-d(t)]\cdot A\cdot\in = s(t)\cdot\pi\cdot r_{eq}^2$$

where the length s(t) of the liquid in capillary is given by relation (21) from [1]:

$$s = \Psi(\text{liq}, \text{pap}, \text{int}) \cdot t^{\alpha}.$$

Relation (2) can be written in terms of parameters Ψ and α, and after some calculations, we can summarize as:

$$\frac{\Im(t)}{\Im_{max}} = \frac{1}{L} \cdot \frac{\pi \cdot r_{eq}^2}{A \cdot \in} \cdot \Psi \cdot t^{\alpha} \tag{3}$$

With relations (1) and (3) we have the ingredients to analyze the plots from Figure 3. In the linear-linear representation, the measured data have been normalized to the highest measured value. Following the algorithm presented in the theoretical part (see reference [1]) we have found:

a) There are time intervals where for both experiments the value for the α parameter obtained from the log-log representation (see Figure 3) is 1. This means that the main transport mechanism of liquid in porous paper is the same in both experiments. A special mention has to be made for the experimental data measured with method (ii). Although it is not proved yet, the none zero value of the water signal for the first 10 sec can be assigned to the moisture that was initially absorbed into paper (just due to the relative humidity of the environment). Then the slow increase for the next 10-20 sec could be assigned to the liquid advance along the fibers before the pore becomes filled.
b) From the plot of $(\Im_0 - \Im(t))/\Im_0$ as a function of t^{α} for method (i) and $\Im(t)/\Im_{max}$ as a function of t^{α} for method (ii) the values of the slopes give indication about the Ψ parameter. For the experiment with water direct in the ATR crystal the slope was found as $1.15\ 10^{-2,}$ while for the method (ii) it amounts $6.61\ 10^{-3}$. From relations (1) and (3) the slopes of the plots are:

$$\text{slope (i)} = \frac{2}{\delta_p} \cdot \frac{\pi \cdot r_{eq}^2}{A} \cdot \Psi$$

$$\text{slope (ii)} = \frac{1}{L} \cdot \frac{\pi \cdot r_{eq}^2}{A \cdot \in} \cdot \Psi \qquad (4)$$

and considering the values experimentally found, we can get:

$$\frac{L}{\delta_p} = \frac{1}{2\in} \cdot \frac{\text{slope (i)}}{\text{slope (ii)}} = 1.7$$

This result is about half from what Mirabella has found in 1983 for porous materials where the solid skeleton remains the same during the experiment [5]. In our situation, the paper skeleton changes when water is absorbed into cellulose fibers and penetrates through the pores. Therefore, we consider this result as validating the two IR-ATR methods for investigating the liquid penetration in porous paper.

4. Using the ATR-IR Spectroscopy to Investigate the Liquid Penetration in Paper

4.1. The Influence of the Paper Relative Humidity on the Liquid Penetration Rate

In the next three sections, experimental data obtained using method (i) of investigation (the liquid is filmed on the ATR crystal and it penetrates in the porous paper placed on top) are analyzed. The influence of the relative humidity from the paper on the α and Ψ parameters is hereunder analyzed. Before the experiment has been made, the paper has been kept (conditioned) for 24 hours in well controlled relative humidity of 15%, 44.3% and 80.5%. The samples have been measured with precautions (the sample transfer in the measurement cell was as short as possible). In order to compare the experimental data we have to make a supposition regarding the internal structure of the paper: it remains the same independently of the relative humidity where it was conditioned. This is a rough assumption. Considering only the water signal, the data are shown in Figure 4a, where the peak height versus time is represented. In the linear- linear representation the slope that reveals the decreasing of the signal in time is smaller for samples with high humidity level. This means that the water front advances with a lower speed when the paper has more water vapors in its structure. The end level of the water ATR signal is also dependent on the initial level of humidity in the paper: it is larger for higher values of relative humidity. This result is expected.

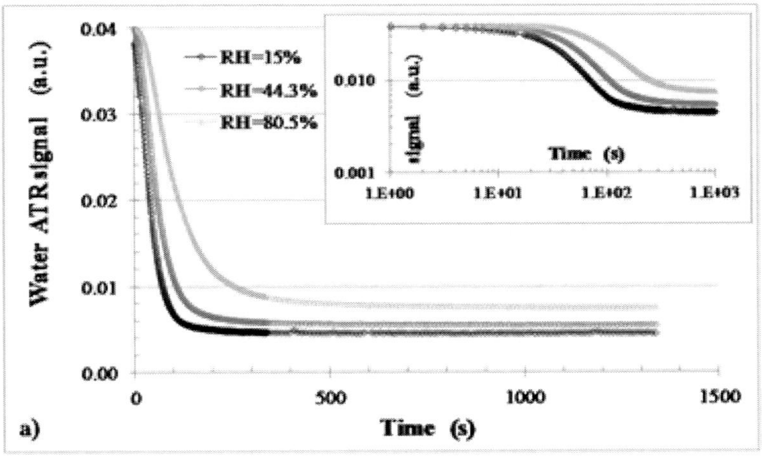

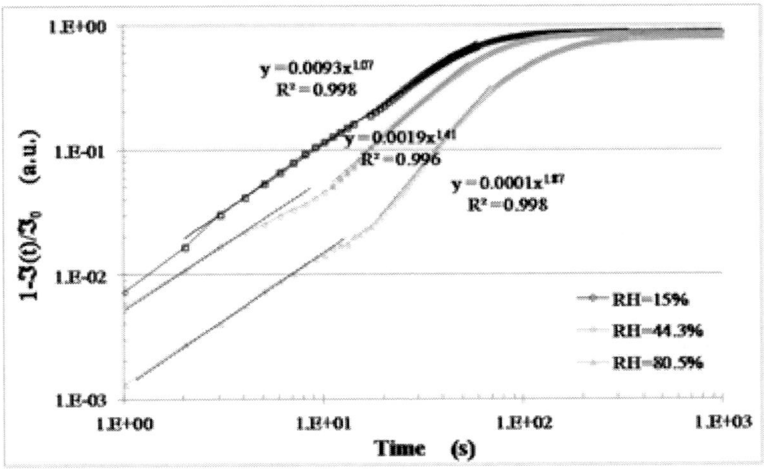

Figure 4. The experimental data for the water penetration in plain paper conditioned at different relative humidity values: a) raw data of the water signal (1639 cm^{-1}) ; b) the representation proposed in previous chapter to reveal the liquid penetration; the values for the α parameter are given, too.

The insert of Figure 4a shows the same data in the log-log scales. Besides the differences in the end levels of the water signal, we note the beginning of decreasing region for each experiment: for samples conditioned at lower RH values is smoother. The α values are obtained from $(\Im_0 - \Im(t))/\Im_0$ versus time plots in the log-log scales. For the measurements at room temperature, according to Figure 4b, the relative humidity has a strong influence on the α values: α=1.04 for RH=15%, α=1.41 for RH=44.3% and α=1.99 for RH=80.3% when t>10s. This means a linear dependency described by the empirical formula:

$$\alpha = 1.4 \cdot RH + 0.84 \tag{5}$$

For the first t<10s the measurements showed α=1.04 for all investigated sets of samples.

Table 1. The slope values of $(\Im_0 - \Im(t))/\Im_0 = f(t^{\alpha})$ plots calculated for the two sets of the α parameter

RH (%)	Slope (α1)	Slope (α2)
15	$9.44\ 10^{-3}$	$9.44\ 10^{-3}$
44.3	$3.91\ 10^{-3}$	$1.74\ 10^{-3}$
80.5	$9.84\ 10^{-4}$	$1.34\ 10^{-4}$

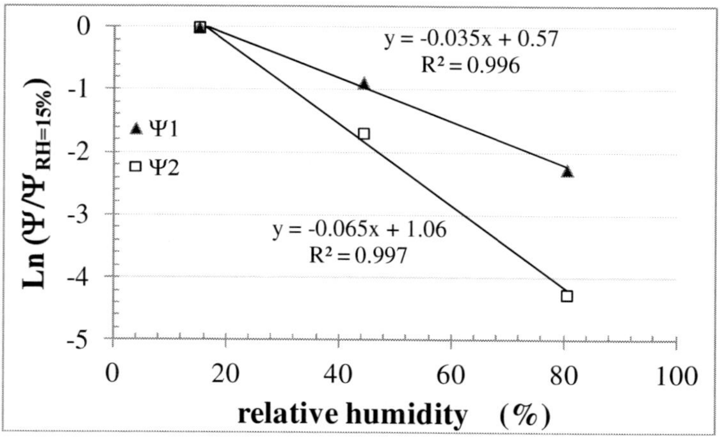

Figure 5. The normalized Ψ values versus the relative humidity of the paper reveal an exponential decreasing.

Having the values for α, the plot $(\Im_0 - \Im(t))/\Im_0 = f(t^\alpha)$ gives information about Ψ via the plot slope. The data obtained for the α values when t<10s and t>10s, respectively, are given in Table 1. According to relation (4), there is a proportionality factor between the plot slope and the Ψ parameter. Considering the same geometry for the measurement and the same porosity for paper, by normalizing the values of "slope (i)" to its value for RH=15%, the influence of the relative humidity on Ψ parameter is obtained: increasing the relative humidity the Ψ value decreases. This is shown in Figure 5 where the exponential dependence for t>10s is revealed, and the result can be written as:

$$\Psi = \Psi_{RH=15\%} \cdot \exp(-0.035 \cdot RH) \tag{6}$$

4.2. Temperature Influence on the Liquid Penetration in Porous Paper

A similar experiment can be carried out varying the temperature of the ATR crystal. Moreover, the paper has been conditioned at the measurement temperature before placing it on top of the liquid layer. The raw data of the measurements at 25°C, 35°C, 40°C, 45°C and 50°C, in terms of absorption peak height assigned to water, are shown in Figure 6a: increasing the temperature the peak height decreases in time at a larger rate. This means that the water penetration rate in the porous paper increases. The plots in log-log scales (see Figure 6b) reveal better the evolution in time of the liquid penetration and the values for the α parameter. Different values have been found for t<10s and for t>10s (shown in Table 2). We note a slight dependence on temperature for the α1 values and a drastically one for the values of α2. Plotting these data in an Arrhenius plot [Log(α)=f(1000/T)], the temperature dependence is suggestively revealed for α2 values as:

$$\alpha(T) = \alpha_0 \cdot \exp\left(-\frac{\Delta E}{k_B T}\right) \tag{7}$$

with ΔE=0.38eV (see Figure 7).

Following the algorithm of this work, the corresponding values for the Ψ parameter can be determined using the representations of $(\Im_0 - \Im(t))/\Im_0 = f(t^\alpha)$ and their slopes. By normalizing the slopes to the slope value obtained for the measurements at 25^{0}C, the influence of the temperature increase on the values of Ψ parameter is revealed. All these data are given in Table 2. An exponential dependence on temperature of Ψ parameter was found and mathematically it can be expressed as:

$$\frac{\Psi_1(T)}{\Psi_1(25C)} \propto \exp(0.14 \cdot T)$$
$$\frac{\Psi_2(T)}{\Psi_2(25C)} \propto \exp(0.22 \cdot T) \tag{8}$$

where T is in Celsius degrees (^{0}C).

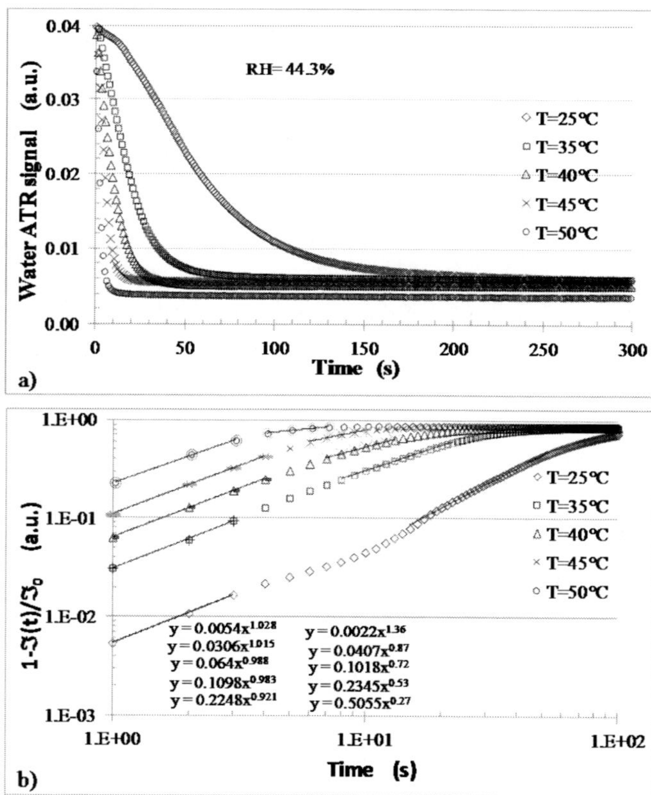

Figure 6. The water peak intensity values versus time for different temperatures values (a) and their representation in log-log scales to obtain the α parameter values (b).

Table 2. The α values presented as a function of the measurements temperature, for t<10s (α1) and t>10s (α2). They have been used to calculate the normalized Ψ values from the slopes of $(\Im_0 - \Im(t))/\Im_0 = f(t^{\alpha})$ plots.

T (C)	α1	α2	slope1	slope2	Ψ1/Ψ1_25C	Ψ2/Ψ2_25C
25	1.028	1.36	0.0048	0.0021	1.00	1.00
35	1.014	0.87	0.0279	0.041	5.81	19.52
40	0.988	0.72	0.0544	0.1005	11.33	47.86
45	0.983	0.53	0.0975	0.2327	20.31	110.81
50	0.958	0.27	0.1795	0.4993	37.39	237.76

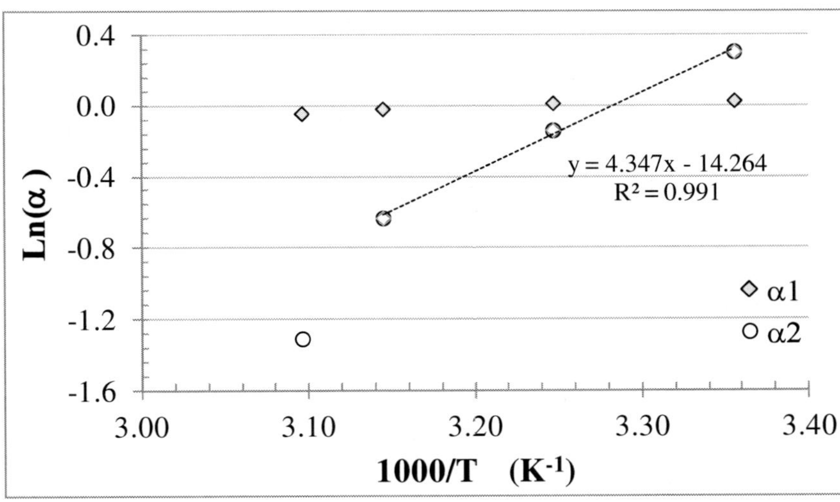

Figure 7. The α values as a function of measurements temperature in an Arrhenius representation.

4.3. Water Penetration in Papers of Different Thickness

Investigating the water penetration in the same type of plain paper but with different thicknesses is interesting. For a better understanding of the obtained results, some details about the paper used in this experiment are necessary; samples of plain paper 80 g/m^2 and 160 g/m^2 have been employed. Their thickness values have measured as: 107 μm and 185.4 μm, respectively. Using the SEM microscopy, information about the so-called 2D porosity is revealed when the paper cross-section is analyzed. This 2D-porosity is the ratio of the area assigned to voids versus the total area of the analyzed cross-section and it could depend on the position where the cross-section is made. In other words, this value is different from the porosity value determined via helium pycnometry and/or mercury porosity measurements. The main results are summarized in Table 3. For the thinner paper, although the raw materials have been the same, the 2D-porosity is larger than the one found for the thicker sample. Both, the pore density and the dimension of pores (maxim pore size), are bigger for the sample of 80 g/m^2. We mention that the porosity value measured via helium pycnometry is 53.7% for the thin paper. Considering the density and the thickness values of the two types of papers, the porosity value was calculated for 160 g/m^2 as 32%. This value fits well in the picture of the 2D porosity obtained via SEM measurements.

The results of the ATR measurements on the signal assigned to water are shown in Figure 8. Plotting the raw data in the lin-lin scale, there are two issues that immediately are observed: i) the ATR signal decreases more rapid when the paper of 160 g/m^2 is used as porous absorber, and ii) the level of the signal at the end of the experiment depends on the paper type.

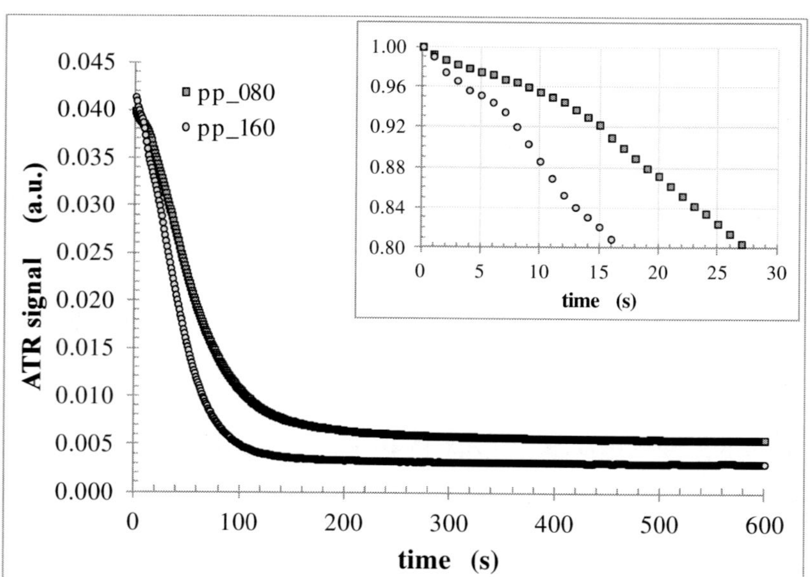

Figure 8. The water characteristic peak versus time when water penetration in plain paper of 80 g/m^2 and 160 g/m^2 is investigated.

In the insert from Figure 8 a detail of the beginning of the experiments is shown; for an easy comparison the ATR data have been normalized to the initial value. The water signal decreased with 5% in the first 5sec when the experiment was made with plain paper of 160 g/m^2 paper. The same percentual decreasing for the experiment with 80 g/m^2 paper occurs in 10.5sec. Also, a knick point is revealed for both experiments in the plots of water ATR signal versus time (after this point the plot slope increases). In other words, the beginning of the experiment is marked by low penetration rate for a limited interval of time.

Table 3. SEM revealed characteristics of the investigated plain paper with different thickness values

plain paper	80 g/m^2	160 g/m^2
2D porosity (%)	13.12	7.40
Pore density (1/µm²)	0.09	0.05
Max Pore Size (µm)	49.4	41.14
3D porosity (%)	53.7	32*

*This value has been calculated, not measured.

Following the algorithm used in the precedent sections, the obtained results are:

a) from the log-log plot of $(\Im_0 - \Im(t))/\Im_0 = f(t)$ the parameter α have been found: 1.42 for paper 80 g/m^2 and 1.08 for paper 160 g/m^2;

b) the slope in the lin-lin plot $(\Im_0 - \Im(t))/\Im_0 = f(t^{\alpha})$ was found as: 1.81 10^{-3} for the thin paper and 9.39 10^{-3} for the thick one; these values are useful to determine the Ψ value following the first relation from (4):

$$\frac{\Psi_{pp-160}}{\Psi_{pp-80}} = \frac{\text{slope}_{pp-160}}{\text{slope}_{pp-80}} \cdot \left(\frac{r_{eq-pp-80}}{r_{eq-pp-160}}\right)^2$$

that results in $\frac{\Psi_{pp-160}}{\Psi_{pp-80}} = 7.39$;

c) the rate of water advancing in paper can be relatively estimated according to our model and for t=30s the penetration speed in plain paper 160 g/m^2 is 132.1% of the penetration speed in the thinner paper. As it can be seen in Figure 8, this value is time dependent and decreases in time.

Therefore, the same liquid in the same environmental conditions of the experiments will penetrate with various rates in plain paper fabricated with different thicknesses. From the fabrication process the parameters of the papers in terms of density, porosity and pore dimensions are different. The results show that the properties of the porous media strongly influence the liquid penetration. Another observation from these experiments is related to the signal values at the end of each experiment: it represent 14.03% from the initial value for paper 80 g/m^2 and 7.41% from the corresponding initial value in the experiment with 160 g/m^2 paper. At the end of the experiment the paper skeleton is in close contact with the ATR crystal and the measured water signal represents the water from the pores. When comparing with the porosity data presented in Table 3, we see that these values fit well. The small difference can be related to the water absorption in the fiber of the paper, and to the error of the method. With some precautions and/or special care, this method can be used to indirectly reveal the porosity of the paper.

5. The Influence of the Liquid Nature: Glycerol versus Water

5.1. Pure Liquids: Glycerol versus Water

Changing the liquid nature, while keeping the same porous paper and experimental conditions, is instructive because the influence of the liquid properties on the two parameters α and Ψ is revealed. For comparison we have chosen water and glycerol; before analyzing the experimental data, we note the differences in some physical properties that are influencing the penetration in a porous material (Table 4). The speed of the liquid advancing front into a porous media, based on two well-known theoretical models, was given by relation (22) in the first part of this work [1]:

$$v = \frac{ds}{dt} = \begin{cases} \sqrt{\dfrac{2 \cdot \sigma \cdot \cos\theta}{r \cdot \rho}} & \text{inertial flow} \\ \sqrt{\dfrac{\sigma \cdot \cos\theta}{8\eta}} \cdot t^{-1/2} & \text{Lucas - Washburn flow} \end{cases} \tag{9}$$

Applying these formulas for water and glycerol in the same type of paper (the same pore effective radius) and considering the contact angle for water 77^0 and for glycerol 70^0 [6] the speed of glycerol advancing front relative to the one of water is:

$$\frac{v_{GLY}}{v_W} = \frac{ds}{dt} = \begin{cases} 89.07\% & \text{inertialflow} \\ 2.37\% & \text{Lucas- Washburn flow} \end{cases} \tag{10}$$

where for the Lucas-Washburn model we have considered the values for t=1sec, for both glycerol and water.

We agree that the discrepancy is huge between the two models. From this point of view we consider our experimental model presented in previous chapter and briefly described hereunder. The measurements for both liquids have been made under the same experimental conditions: 44% relative humidity and 20°C. The signal assigned to water in the IR spectrum is the absorption peak centered on $1639 cm^{-1}$, while for glycerol the absorption peak at $1086 cm^{-1}$ has been considered. The calculation of the penetration depth for this IR wavenumber reveals a value larger with 58.8% for glycerol than the one corresponding to water signal.

Table 4. Surface tension and viscosity for water and glycerol

Liquid	**Surf. tension (mN/m)**	**Viscosity (mPa s)**	**Density (g/cm^3)**
water	73	0.8	1
glycerol	48	1420	1.26

The experimental data for water penetration in porous paper have been analyzed in the previous section. The raw data measured in an experiment to study the glycerol penetration in porous paper are shown in Figure 9.

Comparing with the similar data measured for water, we note the different time scale: for water, the signal decreases towards a constant value in hundreds of seconds, while for glycerol this happens in thousands of seconds. The shape of the glycerol IR-signal in time is similar to that found for the water IR-signal: a) a period (up to 200s) when the signal slightly decreases in time; b) a rapid decreasing region, and c) a slow decreasing region that ends asymptotically constant.

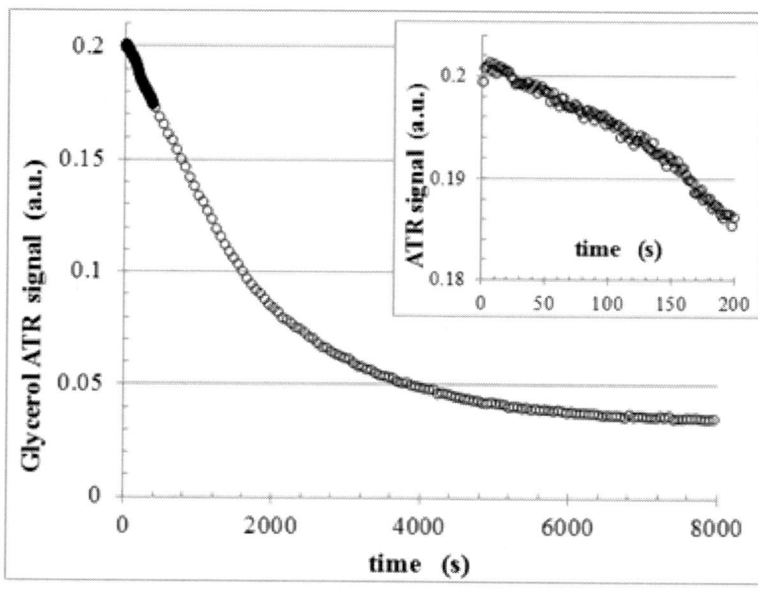

Figure 9. The intensity of the absorption peak centered on $1086 cm^{-1}$ assigned to glycerol, during an IR-ATR experiment to reveal the glycerol penetration in porous paper.

The insert of Figure 9 shows the glycerol penetration in paper in the first 200s. According to this plot, the real variation in signal begins after the first 150s. This interval of time depends on the paper type, but it was found in all investigations we have made. A question that immediately appears is if this delay exists also with water. Because the measured signal for glycerol is different than the signal assigned to water, in Figure 10 we propose the plots of the normalized signal values versus time (the normalization is made against the largest value per each series of measurement). It is clear that the penetration of water in paper (in fact the decreasing of the water signal in the investigated volume) is slower in the first 15 seconds; it is like the system liquid/paper needs a period of "accommodation" prior the liquid advancing in the porous paper proceeds with an increased speed. Therefore, the beginning of the liquid penetration is similar for the both investigated liquids. An interesting observation is that the ATR signal of water decreases with less than 5% in the first 10 sec; after this time, the rate of decreasing is larger. From the glycerol experiment, according to Figure 10 the same 5% decrease of the assigned ATR signal occurs in 150 sec: one order of magnitude longer than the time needed for water signal. We point out that the surface interaction between liquid and paper needs a time interval to be efficient for the liquid transport, and this time is drastically dependent on the liquid properties.

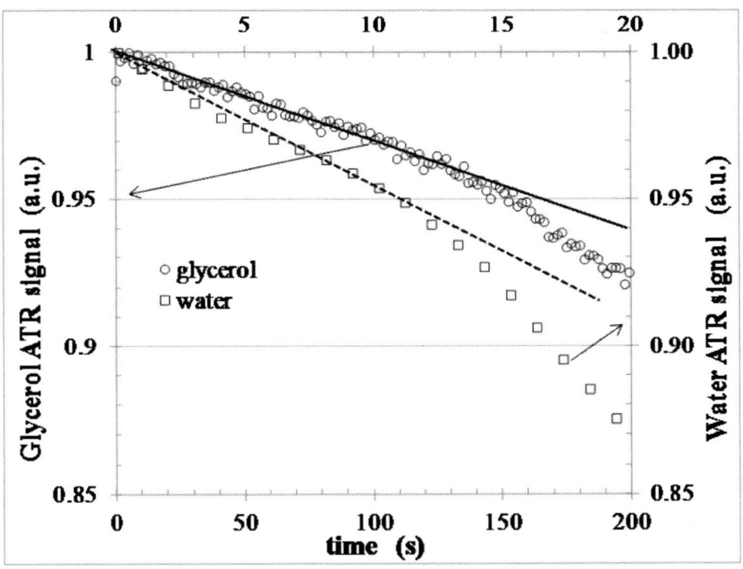

Figure 10. The normalized ATR signal for water and glycerol shows a slower advancing speed at the beginning of the experiment (a comparative study). We note the different time scales for the two experiments.

Figure 11 shows the glycerol penetration in paper according to the model used in this work: $(\Im_0 - \Im(t))/\Im_0 = f(t)$. In the log-log scales representation, the values of parameter α are shown: 1.15 for the first 200s and 0.91 for 200s<t<1800s. We mention that in the first interval of time the errors are large. As it was shown in previous section, the slopes of the $(\Im_0 - \Im(t))/\Im_0 = f(t^{\alpha})$ plots are proportional to the Ψ parameter values. When α=1.15, the slope amounts 1.66 10^{-4}, and for the time interval when α=0.91, it is 6.21 10^{-4}. Using the definition for the plot slope (relation (4)), the Ψ value of the glycerol penetration can be calculated versus its value for water penetration in the same type of paper:

$$\frac{\Psi^g}{\Psi^w} = \frac{\delta_p^g}{\delta_p^w} \cdot \frac{\text{slope(g)}}{\text{slope(w)}} = 0.458$$

This shows a difference in the penetration rate. Indeed, our calculations have revealed an average rate of penetration for glycerol that represents 1.71% from the rate of penetration for water. We mention that the rate of penetration is time dependent. These results are obtained when the pure liquids have been studied.

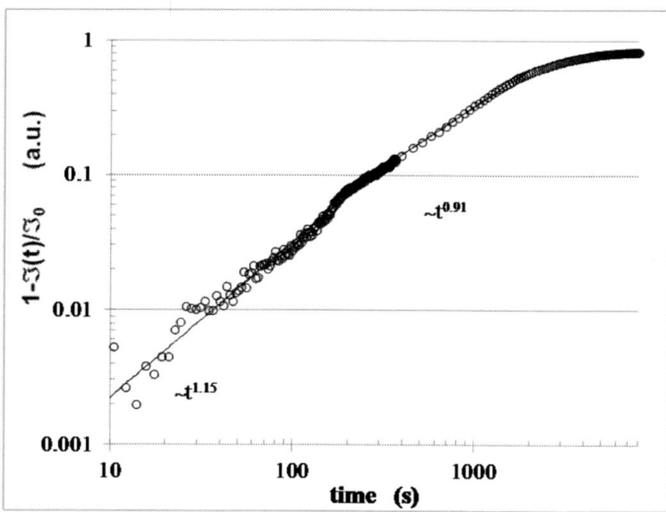

Figure 11. Experimental data for glycerol penetration in plain paper, according to the model here proposed.

5.2. Mixture of Glycerol and Water

In applications many times are used mixtures that contain glycerol and water. In the previous section we have seen the similitude and the differences in the penetration of pure glycerol and water. *But what happened when a mixture of glycerol and water is considered*? It is known that water and glycerol are two liquids perfect miscible in any concentration. This means that the mixture of 50% water + 50% glycerol is homogeneous and in this paragraph we will focus on the time evolution of the ATR signals assigned to water and glycerol measured concomitantly. The paper is the same as in the experiments with pure liquids.

Following the same algorithm as for pure liquids, the results are shown in Figure 12. The intensities of water peak ($1639cm^{-1}$) and of glycerol peak ($1086cm^{-1}$) absorptions are plotted versus time, using different scales. From this representation it seems that the two liquids penetrate into the paper pores with the same rate. When we focus on the beginning of the penetration, the investigated signals diminish with 5% in 14sec for water and 20sec for glycerol. The first observation is that the "accommodation time" for water is almost the same as in the case of pure water (the difference is within the errors of the method), while for glycerol, there is a huge change: from 150sec for pure glycerol to 20sec for this mixture.

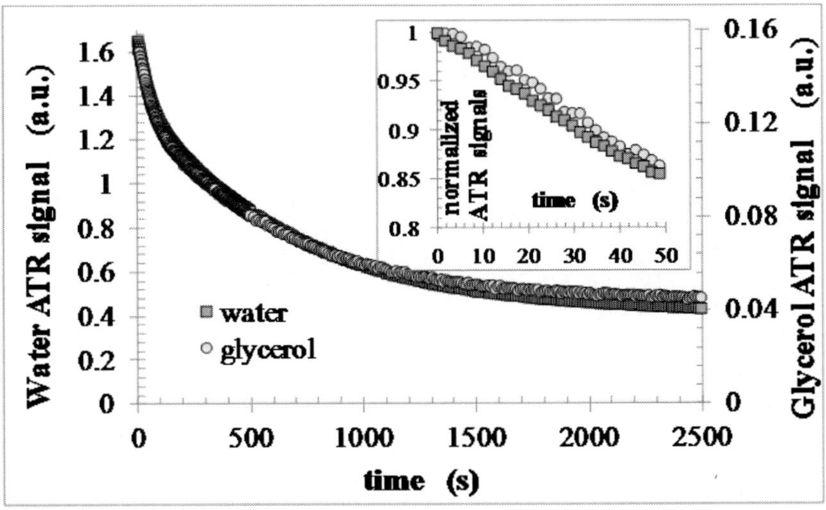

Figure 12. The raw data for the variation in time of the ATR signal of water and glycerol measured simultaneously when liquid penetration of 50% water + 50% glycerol mixture in porous paper has been investigated. We notify the different Y scales utilized for the two signals. In the insert it is shown the evolution of the normalized signals in the first 50sec of the experiment.

The log-log representation of the $(\Im_0 - \Im(t))/\Im_0 = f(t)$ plot is helpful in calculating the α parameter values. As it is shown in Figure 13, at the beginning of the experiment, this parameter amounts 0.95 for water (similar to the case of pure water) and 1.59 for glycerol (is very different than in the case of pure glycerol). This reveals changes in the dynamics of glycerol penetration when water molecules are in the vicinity. In order to determine the second parameter, Ψ, of the model here-used, the representation $(\Im_0 - \Im(t))/\Im_0 = f(t^{\alpha})$ is used. The linear part is well revealed and the result for the two liquids in mixture is: $\Psi^g / \Psi^w = 0.164$. The calculations based on these values show that the advancing rate of glycerol in the first 20sec is 14% from the advancing rate of water. When the calculations are made for a time interval of 125s<t<675s, this ratio is 13.6%. Therefore, in mixtures with water there is an increasing of the penetration rate for glycerol. When the mixture is 50% water + 50% glycerol the increase is close to one order of magnitude.

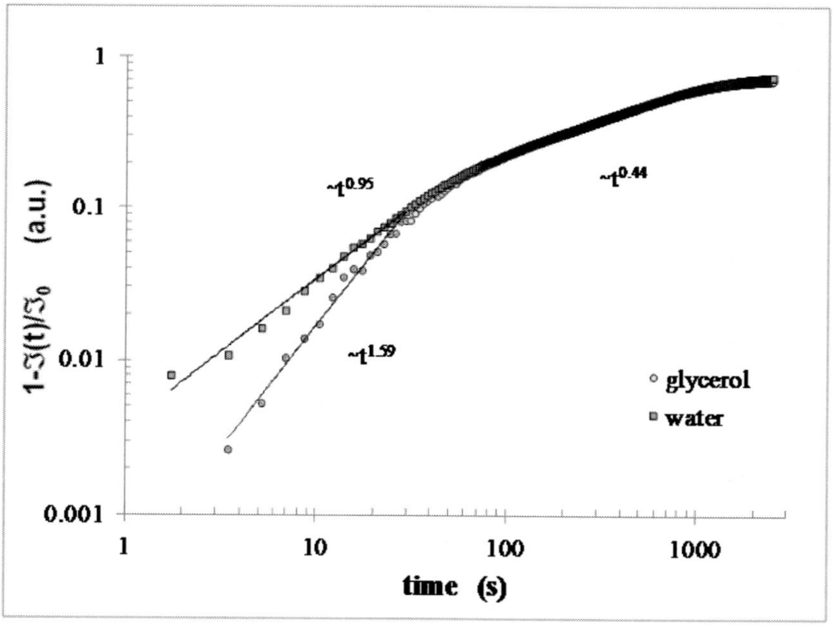

Figure 13. The data from Figure 12 are represented as $(\Im_0 - \Im(t))/\Im_0 = f(t)$ in log – log scales. The values for α parameter are determined. The beginning of the experiment is different for the two liquids: water and glycerol. After 100s our measurements reveal a similar behavior for the two liquids from mixture.

CONCLUSION

Using the IR-ATR measurements and applying the phenomenological model presented in previous chapter we have found that:

a) the two methods here proposed for the IR-ATR measurements to investigate the liquid penetration in porous paper are compatible; however, we mention that when the liquid is sandwiched between the ATR crystal and porous paper, its transport in the paper is indirectly evaluated. It was found that the measured signal belongs to thickness of the investigated material equal to $1.7\delta_p$, where δ_p is the penetration depth of the IR radiation;
b) the liquid penetration in porous paper can be described by two parameters: α, which determines the time dependency, and Ψ that incorporates the dependencies on the characteristics of the liquid and paper, and on the experimental conditions;
c) the α parameter increases linearly with the relative humidity values and exponentially decreases with temperature (for more details see section 4.1 and 4.2);
d) the Ψ parameter is also dependent on the measurement conditions: it exponentially decreases when the relative humidity increases and/or the temperature decreases (for more see the relations (6) and (8));
e) the paper porosity and pore dimensions influence the liquid penetration (small pores imply a larger rate of liquid penetration);
f) the liquid characteristics that determine the interaction with paper drastically influence the rate of penetration (in this work experimental data for water and glycerol are presented);
g) when two miscible liquids are investigated, their penetration rates are different than the ones found in the case of pure liquid investigation.

We consider that the chemistry between the liquid and the paper components plays an important role in the phenomena of liquid penetration through pores and into the pores walls. With the ATR spectroscopy it is difficult to distinguish if the liquid is in pores and/or in the paper fibers. Although there are limitations due to small penetration depth, we consider the IR-ATR spectroscopy a valuable tool to investigate the liquid penetration in porous paper. Based on a good theoretical fundament and using the new technical developments of the IR spectrometry, the ATR methods provide excellent results contributing to better understanding of the processes related to liquid penetration (transport) in porous materials.

REFERENCES

[1] N. Tomozeiu, IR attenuated total reflection – a tool to investigate liquid penetration in paper: theoretical considerations," in *Optical spectroscopy: Technology, Properties and Performances*, Nova Publisher, 2014.
[2] P. Somasundaran, Enciclopedia of Surface and Coloidal Science, vol 7, Taylor & Francis Group, 2006.
[3] Masoodi, Reza and Pillai M.Krishna, Wicking in porous materials: traditional and modern modeling approaches, CRC Press, SBN: 978-1-4398-7432-5, 2012.
[4] M. Denesuk, G. Smith, B. Zelinski, N. Kreidl en D. Uhlmann, "Capillary penetration of liquid droplets into porous materials," *Journal of Colloid and Interface Science,* vol. 158, nr. 1, pp. 114-120, 1993.
[5] F. Mirabella, "Strength of interaction and penetration of infrared radiation for polymer films in internal reflection spectroscopy," *Journal of Polymer Science: Polymer Physics Edition,* vol. 21, nr. 11, pp. 2403-2417, 1983.
[6] K. Mittal, Contact Angle, Wettability and Adhesion, vol. 4, Taylor and Francis Group LLC, 2006, p. 227.

In: Optical Spectroscopy
Editor: Nicolae Tomozeiu

ISBN: 978-1-63321-197-1

Chapter 6

ELECTRONIC SPECTROSCOPY OF DIFFUSE REFLECTION - A PROMISING METHOD FOR QUANTIFYING THE COORDINATION STATES OF THE ATOMS IN THE SUPERFICIAL LAYER OF DISPERSED MATERIALS

Eugene A. Sosnov [*] ***and Anatoly A. Malkov***
Saint-Petersburg State Institute of Technology, St. Petersburg, Russia

ABSTRACT

Given the essential role of solid's surface in the physico-chemical processes, it is important to identify the structure of the superficial layer within the thickness of several nanometers. One of the effective physico-chemical techniques to evaluate the electronic state of the solid materials surface is the Electronic Spectroscopy of Diffuse Reflection (ESDR) in the UV-Vis-spectrum. However, the presence at the surface of several types of structures with different spatial environment makes it difficult to quantify each such structure from the ESDR spectra.

In the following chapter we are offering some possible ways to handle the complex diffuse reflection spectra, which allows to distinguish the individual components for objects with superficial atoms that are simultaneously coordinating in several states, we have analyzed the effect

[*] Corresponding Author address – 26, Moskovskii prosp., St.Petersburg, 190013, Russia.
Email: sosnov@lti-gti.ru.

of nature of the reference sample to the form of the spectral curves and detail of information obtained during the treatment of the spectra.

On the example of the dispersed titanium dioxide we have shown the opportunities of ESDR to examine the state of the superficial layer in the phase transformations anatase - rutile during the anatase heat treatment in the range 200-900°C, and titanium dioxide P25 (Degussa) that incorporates both anatase and rutile.

Keywords: Electronic spectroscopy of diffuse reflection, reference sample, mathematical processing, coordination states

INTRODUCTION

One of the important factors determining the sorption, catalytic and other properties of solids is their phase composition. The significant role of the surface in these physico-chemical processes makes it extremely important to monitor not only the structure of the bulk phase, but also the structure of the superficial layer within the thickness of several nanometers. However, traditionally the information about the phase composition generally refers to the volume of solids and contains no data on the structure of their surface layers. At the same time, structural phase transformation under the heat treatment takes place in the superficial layer at considerably lower temperatures relative to the transformations in the volume [1].

For example, titanium dioxide is known for its chemical stability in acidic and neutral liquids, electrical conductivity, photocatalytic activity, so it is widely used in the manufacture of fillers for composite polymer materials, in dielectric ceramics, pigments, electronic devices, photoelectrochemical cells, photocatalysis, chemical sensors and solar cells, etc. [2, 3]. Thus, depending on the crystalline modification (Anatase, Rutile and Brookite) titanium dioxide shows different physical and chemical properties. At simultaneous presence of amorphous, anatase and rutile component there was increasing of photocatalytic activity of titanium-oxide material compared to the material consisting of only one crystallographic phase [4, 5].

One of the effective physico-chemical techniques to evaluate the coordination state of the atoms at the surface of solids (and, accordingly, to enable the prediction of functional properties of the solid material) is the Electron Spectroscopy of Diffuse Reflectance (ESDR) in the UV- and Visible spectrum. Considering the depth of radiation penetration into solids in electronic spectra, the method can be used to identify the phase composition of the surface layer ~5 nm thick [6]. However, processing of electronic spectra is usually limited by the determination of only one crystalline modification. At the same time on the surface of active materials might be simultaneous presence of several types of structures with different spatial environment of the central atom [7].

THEORETICAL FOUNDATIONS OF ESDR

Consider the electronic structure of solids and manifestations of its features in the optical electronic spectra. Figure 1 shows the total energy scheme, reflecting the specificity of electronic interactions in oxides. This scheme shows the valence *ns-*, *np-* and *(n-1)d*-orbitals of the element forming the oxide, and *2s-*, *2p*-orbitals of oxygen. The collective energy states of the solid particle that form energy layer are represented in the central part. The legitimacy of using molecular orbital theory [8] is proved in case the solid particle are regarded as a "giant macromolecule" (supramolecular structure) [9], which may possess a periodic structure.

In the interaction between valence orbitals of the metal and oxygen atoms in a solid there appears quasi-continuous band states which is responsible for the formation of metal-oxygen bonds $MO^{B}{}_{M\text{–}O}$ (the zone of binding electronic states).

Higher energy is a narrow zone of non-bonding orbitals (Available from participating in chemical bonds, electrons of oxygen are localized on the oxygen atoms, and in the scheme - at non-bonding pseudo-atomic orbitals $MO^{NB}{}_{O}$). Note that the gap between $MO^{NB}{}_{O}$ and $MO^{B}{}_{M\text{–}O}$ showed at the scheme may be absent (wide band of binding states). In this case, the whole continuum is defined as the valence band.

Next is a zone formed by *(n-1)d*-valence orbitals of the metal. Typically, this zone is narrow; its electrons filling degree can be easily determined from the valence configuration of the atom at the formal oxidation state. We must take into account that the *d*-orbitals of the metal split in the crystal field of different symmetry (Figure 2). But in solids the splitted orbitals can overlap without the formation of the gap between the subbands. This process leads to a change in the energy gap between the non-bonding orbitals of oxygen and free *d*-orbitals of the metal, which depends on the coordination environment of metal atom.

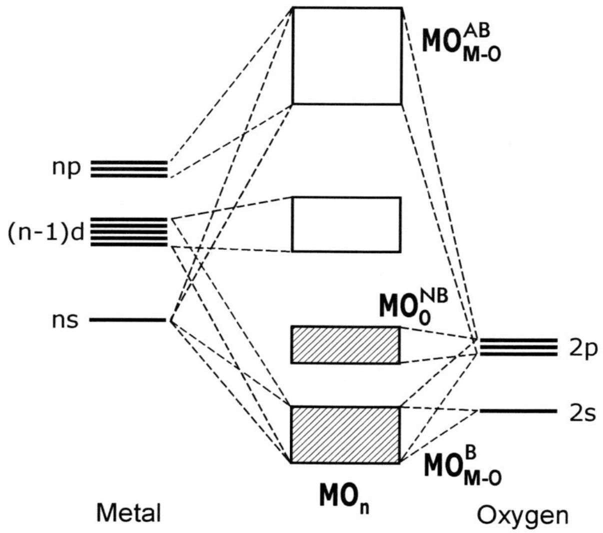

Figure 1. Scheme of the electronic structure of oxides.

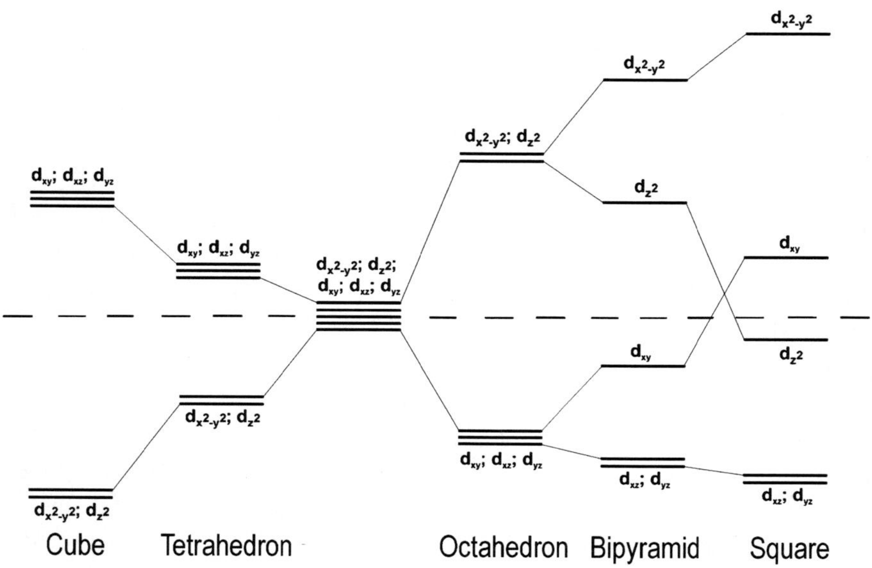

Figure 2. d-Orbital splitting in crystal field of different symmetries.

At the top of the scheme is a zone of free antibonding states $MO^{AB}{}_{M-O}$. Their number corresponds to the number of employed bonding orbitals $MO^{B}{}_{M-O}$.

Short-wavelength region of the optical electronic spectra formed, usually, as a result of electron transitions from nonbonding orbitals of oxygen $MO^{NB}{}_{O}$ to *(n-l)d*-states of metal, which manifest themselves as broad and intense absorption bands at the edge of Visible and near-UV region. The absorption bands edge corresponds to the electrons transition from the highest filled orbitals of the valence band to the lowest unoccupied orbital of the conduction band. The corresponding energy gap is determined in solid-state physics as "forbidden band width", and it is a fundamental characteristic of the solid.

REPRESENTATION OF THE ESDR-SPECTRA

Despite the individual manifestations of the absorption band for structures with different coordination environment [10] (in the case of the simultaneous presence of several different types of coordination structures on the solid's surface) the allocation of their contribution to the spectral transitions is quite difficult [7, 11-13].

The possibilities of various representations of ESDR-spectra will be shown at the example of model two-phase system – pyrogenic titanium dioxide P25 Degussa.

One of the frequently used methods for the estimation of the coordination state of atoms on the surface from electronic diffuse reflectance spectra is the determination of the long-wave fundamental absorption edge by the method of tangents [14, 15] (Figure 3 a). The results obtained this way are compared with the optical forbidden bandwidths for material polymorphs, which allow the structure of the surface of the sample to be qualitatively estimated [16, 17]. Despite the apparent simplicity and reliability of the method, it cannot be used to obtain reliable data on the structure of the surface of materials for two reasons.

1. The position of the absorption band edge determined by the method of tangents substantially depends on the selection of the curve region used for linear extrapolation [12]. For instance, the approximation of the spectral curve of P25 (Figure 3 a) over the range 25,500–24,500 cm^{-1} (approximation validity coefficient R^2 = 0.991) places the absorption band edge at 24,250 cm^{-1} (412.4 nm), whereas spectrum linearization over the range 26,000–24,500 cm^{-1} (R^2 = 0.994) shifts the position of the absorption band edge to longer waves, 24,190 cm^{-1} (413.4 nm).

2. If several structures with different first coordination spheres are present on the surface, the method of tangents usually allows the absorption band edge

for only one type of structures to be determined, most often, for structures with the smallest optical forbidden band width [12]. For our sample, the absorption band edge is situated at 24,190 cm^{-1} (413.4 nm). This gives a 3.00 eV optical forbidden bandwidth, as is characteristic of the rutile TiO_2 polymorph [16], whereas, according to the X-ray data, the P25 sample is two-phase and anatase content in the batch of material is not constant [18] and is within 60–90%.

In addition, the band gap, determined by the method of tangent depends on what kind of representation of the spectrum used for the analysis: as a reflection spectrum ($R_D - \nu$) or transmittance - ($T - \nu$) [10, 16].

Another method for estimating surface structures is the representation of diffuse reflectance spectra in the Gurevich–Kubelka–Munk coordinates (F_R) [19]:

$$F_R = \frac{(1 - R_D)^2}{2R_D} = \frac{\alpha}{S} \approx c, \tag{1}$$

where R_D is the diffuse reflectance and α and S are the effective absorption and scattering coefficient, respectively.

The representation of diffuse reflectance spectra in these coordinates (Figure 3 b) also has serious shortcomings. The disproportion of the sizes of linearized regions that appears because of spectrum restructuring in the Gurevich–Kubelka–Munk coordinates prevents us from the determination of quantitative ratios between surface structures with different coordination [7, 12, 20]. In addition, the position of the approximating straight line as previously depends on the selection of the region of experimental data. In particular, the shape of the spectrum in this representation allows several curve regions linearizable with approximation validity coefficients $R^2>0.99$ to be identified (Figure 3 b, regions I, II, and III), and the selection of regions for approximation can be fairly arbitrary. At the same time, if electronic diffuse reflectance spectra in the Gurevich–Kubelka–Munk coordinates are processed according to [21], where a descending spectrum region (region III) over the range 30–70% of the maximum F_R value is used for approximation, the characteristics of only one of the coordination structures are determined (in our case, the structure with the forbidden band width E_g of 3.48 eV, which can be interpreted as the coordination state of titanium in the anatase polymorph.

The forbidden bandwidth of semiconducting materials can be determined using the representation of optical spectra in the coordinates [22]

$$(\alpha h\nu)^m \propto (h\nu - E_g), \tag{2}$$

where α is the coefficient of radiation absorption by the sample, cm^{-1}; hν is the quantum energy, eV; and m is the coefficient depending on the type of electron transitions from the valence to conduction band. For allowed (the summed momentum of electron spins does not change) direct transitions, m = 2, and, for indirect transitions, m = 1/2.

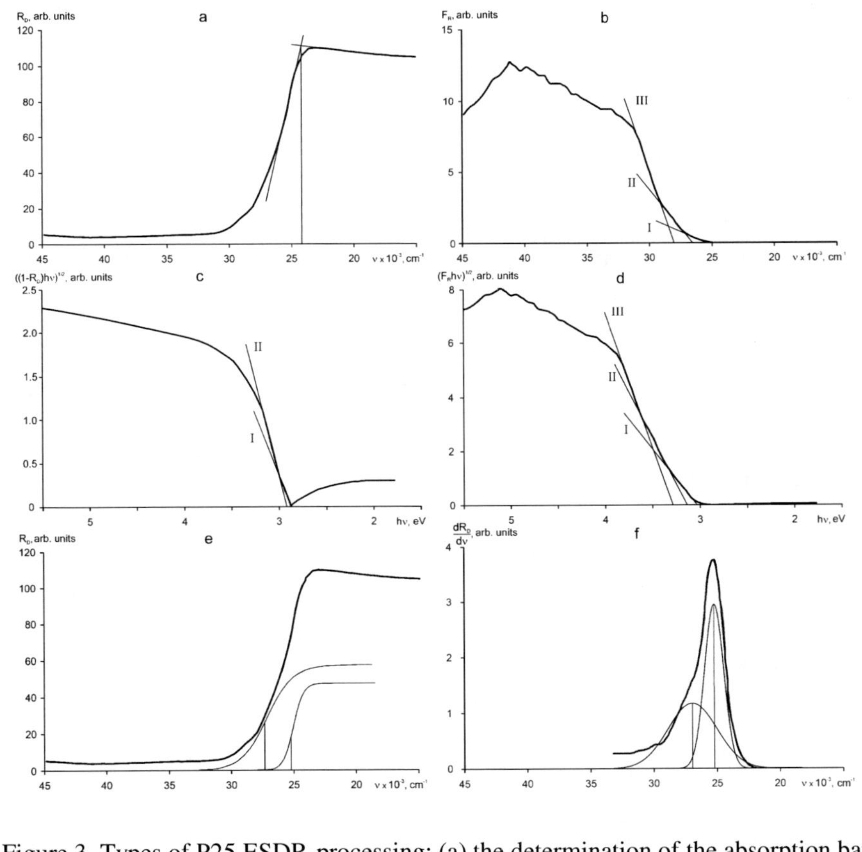

Figure 3. Types of P25 ESDR-processing: (a) the determination of the absorption band edge by the method of tangents, (b) representation in the Gurevich–Kubelka–Munk coordinates, representation in the (c) $((1-R_D)h\nu)^{1/2} - h\nu$ and (d) $(F_R h\nu)^{1/2} - h\nu$ coordinates, (e) spectrum separation into components described by the Fermi–Dirac distribution, and (f) processing of the results of the instrumental differentiation of ESDR.

If α cannot be determined directly, its normalized calculated values can be used [23],

$$\alpha = 1 - (T + R), \tag{3}$$

where T is the normalized thin film transmission and R is the normalized thin film reflectance. With electronic diffuse reflectance spectra, representation (3) becomes

$$\alpha = 1 - R_D, \tag{4}$$

which introduces determination error E_g because the transmission spectrum of the material cannot be measured.

Since the Gurevich–Kubelka–Munk representation is proportional to the absorption coefficient, it was suggested in [21, 24, 25] that α in (2) could be replaced with the spectrum representation in the Gurevich–Kubelka–Munk coordinates, $(F_R h\nu)^m \propto (h\nu - E_g)$. Because the band gap in TiO_2 was determined by indirect electron transitions in Figure 3 c & d shows the spectra of titanium dioxide P25 in coordinates $((1-R_D)h\nu)^{1/2} - h\nu$ and $(F_R h\nu)^{1/2} - h\nu$, respectively. At the same time, these methods for spectral data representing have the same shortcomings as the representation of spectra in the Gurevich–Kubelka–Munk coordinates. In the spectrum also can be arbitrarily allocated some regions for a linear approximation with a reliability coefficient of $R^2 > 0.995$. These procedures also cannot be used to describe the structure and determine the anatase/rutile quantitative ratio on the surface of the P25 sample.

Methods of Separation the ESDR-Spectrum into Components

In earlier studies [11-13] we were invited to consider the edge of fundamental absorption spectra not from the optics standpoint (i.e., radiation interaction with an infinitely thick scattering layer of closely packed disperse medium particles [26]), as well as a quantum phenomenon - the probability finding an electron with energy E_i close to the quantum transition point (E_0).

The probability of the transition is described by the Fermi–Dirac distribution of particles with half-integral spins [27] (Figure 4),

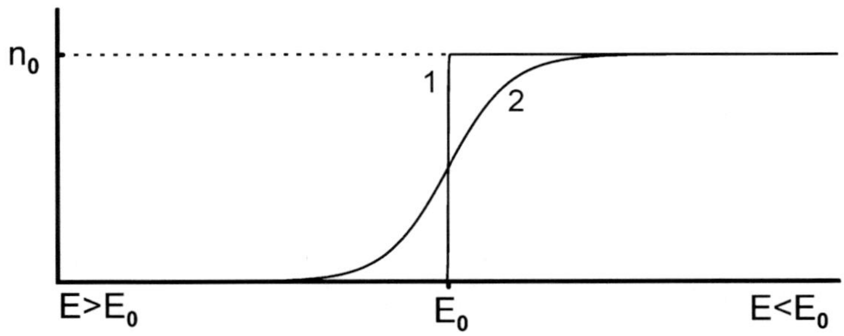

Figure 4. Fermi–Dirac distribution curve at T = (1) 0 K and (2) > 0 K.

$$n_i = \frac{n_0}{1 + \exp((E_i - E_0)/kT)}, \tag{5}$$

where n_i the number of carriers have overcome the band gap; n_0 is the number of mobile carriers; k is the Boltzmann constant; and T is the temperature, K.

At T = 0 K, we then observe a jumptype quantum transition at the $E_i = E_0$ point (Fig. 4, curve 1). At T > 0 K, the quantum transition becomes smeared because of electron thermal energy, and it acquires the form of an S-shaped curve (Fig. 4, curve 2).

We can apply such a representation to the nature of electronic diffuse reflectance spectra and, in the presence of several (j) types of surface structures with different coordination states on the TiO_2 surface, describe each transition by the Fermi–Dirac distribution in the form

$$I_{ji} = \frac{I_{0j}}{1 + \exp((E_i - E_{0j})/\beta_j)}, \tag{6}$$

where I_{0j} is the transition intensity for the j-th type of surface structures, I_{ij} is the contribution of j-th type structures at the i-th point of the spectrum, E_{0j} is the quantum transition energy for surface structures of the j-th type, and β_j is the coefficient characterizing smearing of the transition band.

The absorption band edge in electronic diffuse reflectance spectra can be represented in the form

$$I_i = \sum_j I_{ji} = \sum_j \frac{I_{0j}}{1 + \exp((E_i - E_{0j}) / \beta_j)}, \tag{7}$$

and mathematical processing can be applied to separate such overlapping spectral transitions (Figure 3 e).

The ideal model structure shows the real titanium-oxide complexes on the TiO_2 surface, although they have similar coordination structures, they can be characterized by somewhat different lengths and angles of valence bonds. This can be caused by the differences in the electron density on oxygen atoms situated in the volume and at the boundary with air. Such distortions of the polyhedron structure should influence the shape of the distribution curve and increase smearing of the quantum transition region above temperature-induced broadening ($\beta > kT$, where kT is ~0.025 eV at T = 293 K). Actually, the β coefficient determined from the Fermi–Dirac distribution characterizes the degree of temperature and configurational disordering of surface titanium-oxide structures.

It should be noted that the literature clearly separate "electronic" and "optical" band gap of semiconductors [22, 28]. The proposed method of ESDR-spectra separation into components described by Fermi-Dirac distribution, gives the energy characteristics of surface structures, comparable to the "electronic" band gap of semiconductors.

To estimate the influence of scanning conditions on the accuracy of the determination to the characteristics of the spectral bands of interest, a series of electronic diffuse reflectance spectra were obtained for P25 sample using different scan steps (from 0.1 to 10 nm). The dependence of calculation results on the scan step is shown in Figure 5.

Thus, scanning step changes from 0.1 to 2 nm do not cause significant changes in the determination of band characteristics, and deviations of the I_0 and β values do not exceed 3%. An increase in the scanning step to 10 nm considerably increases the errors in these values, which can be as large as 20–25%. The error in the position of the middle of the electronic transition (E_0) is substantially smaller and does not exceed 1% over the whole range of scanning step values under consideration.

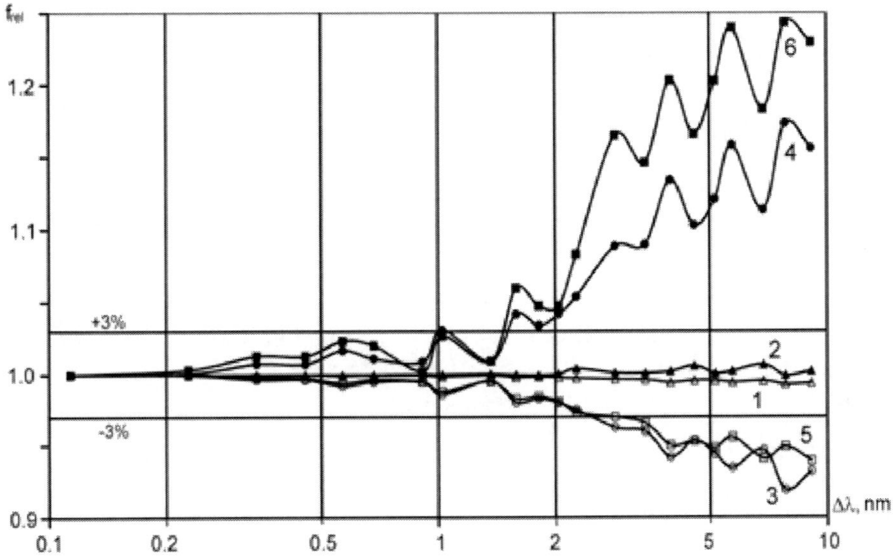

Figure 5. Influence of spectrum scanning step on the relative (1, 2) E_0, (3, 4) I_0, and (5, 6) β values of the distinguished spectral bands. The curves characterize bands (1, 3, 5) 1 and (2, 4, 6) 2 of P25.

In recent years, the exact spectral curve peak positions have been determined using the first derivatives of spectral functions [29, 30]. Overlap of bands related to different coordination structures is, however, possible for differential curves (Figure 3 f), which requires an additional mathematical processing for the separation of peaks. We used the nonlinear regression method to decompose the curve of the instrumental differentiation of the spectrum of P25 into Gauss components,

$$\left(\frac{dI}{d\nu}\right)_i = \sum_j \frac{A_j}{w_j\sqrt{\pi/2}} \exp\left(-2\left(\frac{E_i - E_{0j}}{w_j}\right)^2\right), \qquad (8)$$

The results are given in Table 1.

A comparison of the results of P25 electronic diffuse reflectance spectrum processing obtained using various procedures (Table 1) shows that the results obtained by graphic methods (the determination of the absorption band edge and linearization of spectrum regions in the Gurevich–Kubelka–Munk, $((1-R_D)h\nu)^{1/2}$ or $(F_R h\nu)^{1/2}$ representation) cannot be used to describe not only the quantitative but also qualitative composition of surface structures. At the same time, the separation of spectra into components described by the Fermi–Dirac distribution and the separation of the instrumental differentiation curve into components described by the Gauss distribution allows not only the structure of coordination polyhedra but also the anatase/rutile ratio on the surface of the material to be determined. The difference between the data obtained by these methods is small (within measurement errors of 1–2%).

Table 1. Optical forbidden bandwidths and anatase and rutile contents (x_A and x_R) in the surface layer of P25 pyrogenic titanium dioxide obtained as a result of electronic diffuse reflectance spectrum processing by various methods

Method	Anatase		Rutile	
	x_A, %	E_0, eV	x_R, %	E_0, eV
According to processing by the method of tangents from the absorption band edge	-	-	-	3.00
Gurevich–Kubelka–Munk	-	3.48	-	-
$((1-R_D)h\nu)^{1/2}$	-	-	-	2.87
$(F_R h\nu)^{1/2}$	-	3.29	-	-
Spectrum separation into components according to Eq. (7)	54.8 *	3.39	45.2 *	3.13
Instrumental differentiation curve separation into Gauss components according to Eq. (8)	53.3 **	3.34	46.7 **	3.13

* Determined from the intensity (I, %) of the transition.

** Determined from the area bounded by the Gauss curve.

ROLE OF REFERENCE SAMPLE

For ESDR-spectra preparing it is possible to use spectrophotometers of different manufacturers and different brands equipped by an integrating sphere. Irrespective of the design features of the device (one- or two optical channel) reflectance measurements performed with respect to the optical reference materials, in which using the standard reflection materials (MgO, $BaSO_4$, polytetrafluorethylen (PTFE), also commercially known as Halon, Spectralon (Labsphere Inc.), Zenith Optowhite (Sphere Optics), ets.) [31], calcined pyrogenic silica (aerosil) [32], the white pigment materials (TiO_2, ZrO_2), and in the case studies of thin coatings supported on an inert carrier -

source matrix. At the same time, the reference material itself can have an absorption band in the region of spectral transitions of coordination complexes and thereby may influence the shape of the spectrum.

Assessing the impact of reference sample at the results of spectral measurements was spended on example of silica gel coated according to the method of Molecular Layering method (ML) [9, 33] (also known as atomic layer epitaxy (ALE) or atomic layer deposition (ALD)) by titanium-oxide structures with different coordination environments. The choice of titanium as an element-modifier was due to the fact that the Ti^{4+} has $3d^0$-configuration (the term of basic state is $^1S_1$), and due to the lack of free electrons - term of state in the crystal fields of any symmetry - $^1A_{1g}$. Therefore ESDR-spectra of titanium-oxide structures are not complicated by the presence of absorption bands due to the splitting of the atom-modifier in various coordination polyhedrons and crystal fields of different symmetry [10].

As the carrier we used a fine-pored silica produced by Acros Organics (S = 470 m^2/g, d_{pore} = 6 nm). Synthesis of titanium-oxide nanostructures on the carrier's surface by ML method comprises the following successive stages of carrying out:

1. Heat treatment of the original matrix;
2. Chemisorption of $TiCl_4$;
3. Removal of halide excess and gaseous products of reaction;
4. The vapor phase hydrolysis;
5. Removal of H_2O excess and gaseous products of reaction;
6. Heat treatment of the modified matrix,

where in steps (1) and (6) defining a surface concentration of hydroxyl groups (both involved in the chemisorption reaction and in the formation of coordination complexes [34]) necessarily precede the step of chemisorption of titanium tetrachloride. It has been shown [35] that on the surface of fine-pored silica gel during its sequential processing by $TiCl_4$ and H_2O vapors the titanium-oxide structures were building up, and polyhedrons with different spatial environment of the atom-modifier formed on the matrix surface have depended on the temperature conditions of synthesis.

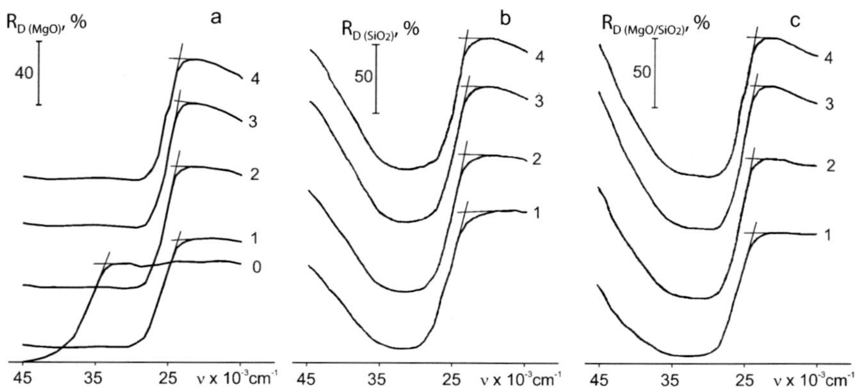

Figure 6. ESDR-spectra of initial silica gel (0) and silica after 1-4 cycles of ML synthesis (1-4) obtained with (a) MgO and (b) initial silica as reference, and (c) spectra recalculated on diffuse reflectance of initial matrix according eq. (9).

Figure 6a shows the ESDR-spectra of initial and modified silica gel obtained with MgO reference sample. The absorption band edge of silica gel, defined by the tangent method, is at 33,600 cm^{-1} (Table 2). In the spectral region of 33,000–15,000 cm^{-1}, where the absorption edge of titanium-oxide structures on the surface (23,850–23,600 cm^{-1}, see Table 2) was observed, the R_D value of the initial matrix obtained with MgO reference is practically constant. However, the presence of the absorption band in the initial matrix in the 45,000–33,600 cm^{-1} can affect the shape of spectra and results of their different mathematical processing. To determine the influence of the reference sample, the samples of modified silica obtained by using MgO as a standard, compared with the spectral data:

- Obtained by using he initial matrix as a reference.
- Obtained with MgO as reference, and then spectra recalculated relative to the original matrix in the entire spectral range by the formula

$$R_{D\,(MgO/SiO_2)} = \frac{R_{D\,(MgO)}}{R_{D\,SiO_2}} \times 100\%\,, \tag{9}$$

where $R_{D\,(MgO/SiO2)}$ – diffuse reflectance of modified sample obtained with MgO as reference and recalculated on diffuse reflectance of initial matrix, %;

$R_{D\ (MgO)}$ – diffuse reflectance of modified silica with MgO as reference, %; $R_{D\ SiO2}$ – diffuse reflectance of initial matrix with MgO as reference, %.

Table 2. The absorption band edge measured by the method of tangents in ESDR-spectra of initial and modified silica gel

Sample	Reference sample								
	MgO			SiO_2			MgO/SiO_2 *		
	nm	cm^{-1}	eV	nm	cm^{-1}	eV	nm	cm^{-1}	eV
initial SiO_2 matrix	297	33,700	4.18	–	–	–	–	–	–
SiO_2-1Ti **	424	23,600	2.93	426	23,500	2.91	424	23,600	2.93
SiO_2-2Ti	419	23,850	2.96	422	23,700	2.94	420	23,800	2.95
SiO_2-3Ti	424	23,600	2.93	427	23,400	2.90	424	23,600	2.93
SiO_2-4Ti	424	23,600	2.93	426	23,500	2.91	426	23,450	2.91

* From recalculated spectra.

** Modified silica (SiO_2-nTi), where n is quantity of cycles of synthesis.

The spectra obtained when shooting relatively SiO_2 (Figure 6 b) and recalculated with respect to the initial matrix (Figure 6 c) are almost identical in shape and position of the absorption band edge. In both cases, in the region of 30,000–15,000 cm^{-1} the reflectance increase was marked, compared with the spectra obtained with MgO as reference. In the range of 45,000–35,000 cm^{-1} an increase in reflectance (Figures 6 b & c) was caused by the absorption in this range of wave numbers in the spectrum of the initial matrix (see Figure 6 a, curve 0). Regardless of whether you are using MgO or SiO_2 as a reference the position of the absorption band edge in the spectra (Table 2) are similar and are in a region characteristic for anatase and rutile. This fact indicates that the coordination state of titanium atoms in the superficial layer is similar to the coordination environment of the metal in bulk crystalline phases.

In differential ESDR-spectra of silicas can see that on the spectrum of the initial silica exhibits a maximum at 35,700 cm^{-1} (Figure 7, curve 0) that was absent at the differential spectra of modified silica gels obtained with MgO as reference. In the range of 30,000–23,000 cm^{-1} in the differential spectra of titanium-containing samples (Figure 7, curves 1-4 and 1'-4 ') there are two peaks with maxima close to the same values for anatase and rutile (Table 3). For the titanium-containing sample obtained after the one cycle of modification, the intensity maxima of both peaks almost equal. With increasing titanium content in the samples, regardless of the choice of standard shooting, there is a tendency to an increase in the intensity ratio of the maximum corresponding to rutile, compared to the same parameter in anatase.

Thus, we can conclude that with the increasing titanium-oxide structures occurs increase in the proportion of titanium atoms having less distorted (rutile-like) coordination environment. It was found that the standard selection could influence the shape of ESDR-spectra and spectra's differential representations.

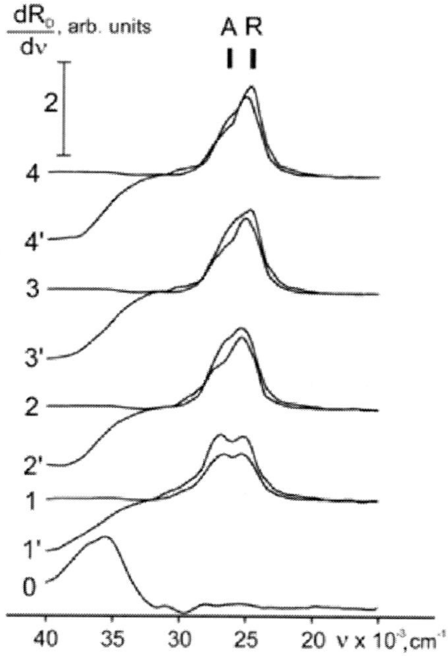

Figure 7. The differential ESDR-spectra of initial (0) and modified silica after 1-4 cycles of ML synthesis obtained with MgO (0, 1-4) and initial silica (1'-4') as reference. A, R - position of maxima in spectra of anatase and rutile, respectively.

Table 3. Position and the intensity ratio (I_A/I_R) of maxima on differential curves of modified silica gel

Sample	Reference sample								
	MgO			SiO_2			MgO/SiO_2		
	I_A, cm^{-1}	I_R, cm^{-1}	I_A/I_R	I_A, cm^{-1}	I_R, cm^{-1}	I_A/I_R	I_A, cm^{-1}	I_R, cm^{-1}	I_A/I_R
Anatase	26,900	–	–	–	–	–	–	–	–
Rutile	–	23,980	–	–	–	–	–	–	–
SiO_2-1Ti	26,700	25,300	1.00	26,900	25,200	1.04	26,800	25,300	1.03
SiO_2-2Ti	27,100	25,400	0.60	26,600	25,400	0.85	27,200	25,400	0.60
SiO_2-3Ti	26,600	25,100	0.65	26,100	24,700	0.84	26,800	25,100	0.57
SiO_2-4Ti	26,400	25,100	0.60	26,300	24,800	0.65	26,400	24,900	0.56

However, if the region of own absorption bands of the reference sample doesn't match the ambit of spectral transitions of the studied surface structures, the use of reference materials is possible. At the same time it should be noted that the use of reference sample with their own absorption bands might introduce error in the quantitative results in separation of curve of hardware differentiation on Gaussian components due to the asymmetry of the background line (see Figure 7, curves 1'-4').

ESDR VS XRD

Because the surface of solid is largely determined the physico-chemical, catalytic properties of material, etc. the sensitivity of ESDR-method to detect the change of surface composition of the material compared with the capabilities of X-Ray Diffraction were evaluated. For a comparison we use the laboratory sample of nanocrystalline TiO_2 (AK-1, S = 133 m^2/g) obtained by the controlled hydrolysis of $TiCl_4$ followed by drying and heat treatment at 200°C.

Consider the effect of heat treatment in the range 200–900°C on the structural transformations of titanium dioxide AK-1. X-ray analysis (the method normally used to characterize crystalline materials) of the samples subjected to thermal annealing up to 600°C, shows (Figure 8) the presence of only anatase modification of TiO_2. During the treatment at temperatures from 200 to 600°C a certain increase of the crystallite size of the anatase phase (Table 4) caused by crystallization of the amorphous component of TiO_2 was observed. In samples annealed at 700°C, along with the increasing intensity of the anatase reflections, indicating further crystallization of the amorphous component of dioxide, the appearing of small-intensive peaks related to rutile was already observed. And only in samples subjected to thermal annealing at 800°C, we can see a decrease in intensity of the reflection, which characterizes the anatase phase, and a significant increase in peaks related to rutile. A sample heat-treated at 900°C, according to XRD data consists exclusively of the rutile modification of TiO_2.

Spectra diffuse reflection AK-1 after heat treatments at various temperatures are presented in Figure 9. To obtain quantitative data characterizing the titanium coordination station the surface of its dioxide, we isolated the region of the long-wave absorption band edge from the electronic diffuse reflectance spectra (from the minimum to maximum R_D value). We assumed that the superposition of quantum transitions described by distribution (7) was observed in this region. The spectral regions were divided into components by the gradient descent method; the results are given in Table 5. If the sample contains coordination polyhedra of only one type (with similar configuration), mathematical processing reduces the intensity of the second band to zero. Simultaneously using nonlinear regression was conducted decomposition curves of hardware differentiation of the spectra of the samples in the Gaussian components.

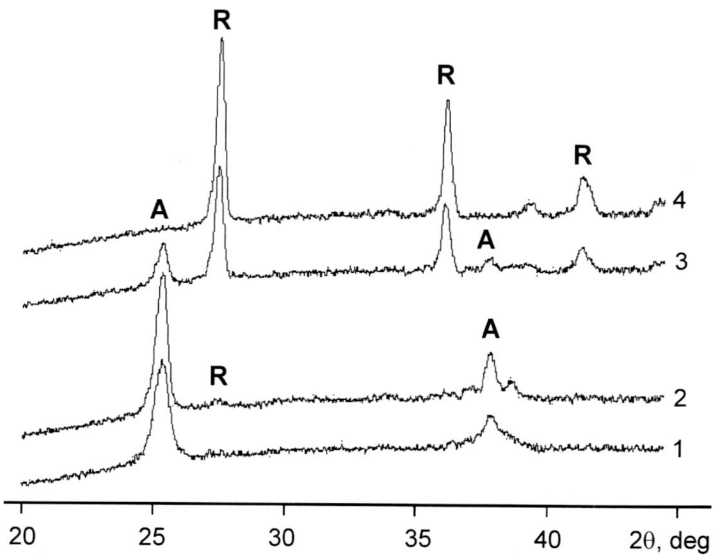

Figure 8. X-Ray diffraction on heat-treated titanium dioxide AK-1 at (1) 200–600, (2) 700, (3) 800 and (4) 900°C. A, R - position of peaks of anatase and rutile, respectively.

Table 4. The results of XRD studies of heated titanium dioxide AK-1

Temperature of heat treatment, °C	Anatase			Rutile			XRD, A/R, %
	θ, deg	d/n, nm	D_{CSR}, nm*	θ, deg	d/n, nm	D_{CSR}, nm	
200	12.52	0.356	16.3	–	–	–	100 / 0
600	12.52	0.356	20.4	–	–	–	100 / 0
700	12.52	0.356	20.9	13.75	0.324	13.6	92 / 8
800	12.52	0.356	20.4	13.74	0.325	27.1	23 / 77
900	–	–	–	13.75	0.324	40.7	0 / 100

* Size of coherent scattering region.

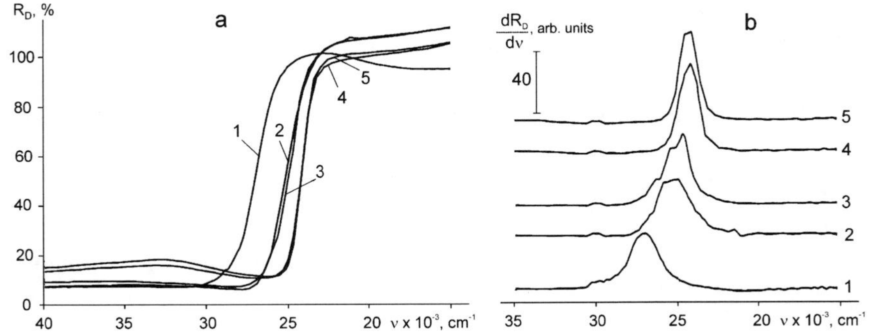

Figure 9. ESDR-spectra (a) and differential ESDR-spectra (b) of the titanium dioxide AK-1 heat-treated at (1) 200, (2) 600, (3) 700, (4) 800, and (5) 900°C.

Considering the results presented in Table 5, if λ or E_0 are fundamental in character and characterize the configuration state of the titanium atom, the absolute band intensities (I_0) obtained as a result of mathematical processing depends on the reference sample. Thus, only the relative intensities ($I_j = I_{0j} / \sum_j I_{0j}$) are characteristic. The determination of quantitative characteristics from the data obtained in the instrumental differentiation of the spectra requires calculations of the areas bounded by Gauss curves.

Comparing the results obtained by separation of ESDR-spectrum into components, described by Fermi-Dirac distribution, and results of separation of hardware differentiation spectra into the Gaussian curves, it is sufficiently close quantitative characteristics obtained by both methods. At the same time, it is shown [36] that the expansion curves of hardware differentiation have less sensitivity especially at low intensities of spectral transitions (see Table 5, T = 200°C). This shortcoming is caused by an error of fixing the actual values of the first derivative due to the oscillation of luminous flux devices. Therefore, the further consideration of the results of the spectral data analysis will be based on the separation of initial spectra.

Mathematical processing of the spectrum of treated at 200°C AK-1, showed (Table 5) that at the specimen's surface the titanium atoms in two coordination states are present at the same time: distorted octahedral which corresponding to anatase crystalline phase (E_0 = 3.34 eV), and in polyhedrons with a maximum absorption band at 3.68–3.70 eV (depending on the type of spectral data), which is typically associated with tetrahedrally coordinated titanium atom [16]. In X-ray diffraction studies the information about structures with tetrahedral coordination titanium, which do not form a separate crystalline phase of TiO_2, can be obtained only indirectly by the change of anatase crystallite size during the heat treatment (Table 4).

Table 5. Results of ESDR-spectra processing of titanium dioxide samples

Temperature of heat treatment, °C	Band 1						Band 2						Method of tangents from the absorption band edge	
	Instrumental differentiation curve separation into Gauss components			Spectrum separation into components according to Eq. (7)			Instrumental differentiation curve separation into Gauss components			Spectrum separation into components according to Eq. (7)				
	λ, nm	E_0, eV	I, %	λ, nm	E_0, eV	I, %	λ, nm	E_0, eV	I, %	λ, nm	E_0, eV	I, %	λ, nm	E_0, eV
200	337	3.68	7.2	334	3.70	3.1	372	3.34	92.8	372	3.34	96.9	380	3.26
600	401	3.09	91.3	395	3.14	91.3	409	3.03	8.7	413	3.00	8.7	418	2.97
700	402	3.08	59.7	402	3.08	60.5	410	3.02	40.3	413	3.00	39.5	418	2.97
800	--	--	--	--	--	--	414	2.99	100	415	2.99	100	420	2.95
900	--	--	--	--	--	--	414	2.99	100	415	2.99	100	420	2.95

With increasing heat treatment temperature of the sample the disappearance of tetrahedral titanium-oxide structures and sharp transition polyhedra structure from anatase-like to rutile is observed. The titanium atoms in anatase and rutile have an octahedral coordination environment of oxygen so the alignment of Ti-O lengths and bond angles ∠O-Ti-O (and, consequently, increasing the symmetry of coordination polyhedron of titanium atoms), the decrease of width band gap of material (from 3.3–3.4 to 2.9–3.1 eV) observed.

So, according ESDR in the samples annealed at 600°C and higher, the titanium atoms in the superficial layer are preferably in the coordination state which is characteristic for rutile modification, whereas according to the XRD results, after the heat treatment at 700°C sample composed of 98 % anatase, at 800°C at 23 % and only the sample calcined at 900°C corresponds to the pure rutile phase.

Thus, unlike the XRD data, already at 600°C the coordination environment of metal atoms at the surface of the sample does not correspond to the atom's coordination in the crystalline phase of titanium dioxide. Similar phenomena were found in the analysis of commercial titania products - anatase and rutiles different manufacturers [36]. Cause discrepancies between the results of spectroscopic and XRD-studies is the mechanism of the anatase-to-rutile phase transition which begins with the grain surface. As a result we can see the advancing phase transformation of anatase to rutile in the surface layer compared with the volume. Anatase core, which does not affect the coordination state of titanium on the particle surface, may be in the center of the particle.

Conclusion

Thus mathematical processing of the ESDR -spectra allows to perform with sufficient accuracy both qualitative and quantitative evaluation of coordination state of atoms on the surface of solids. Proposed approaches of ESDR using in combination with the methods of studying the solids structure allow to organize a comparative analysis of physical and chemical transformations that occur in the superficial layer and in the volume of solid materials, and to predict the nature of the change and hence the functional properties of solids.

References

[1] Tret'yakov, Yu. D.; Putlyaev, V. I. *Introduction to the Chemistry of Solid Phase Reactions;* Nauka: Moscow, 2006; 400 p. [in Russian].

[2] Kiselev, V. F.; Krylov, O. V. *Adsorption Processes on Semiconductor and Dielectric Surfaces I*; Springer-Verlag: Berlin & Heidelberg, 1985; 287 p.

[3] Shabanova, N. A.; Popov, V. V.; & Sarkisov, P. D. *Chemistry and Technology of Nanodispersive Oxides*; Akademkniga: Moscow, 2006; 309 p. [in Russian].

[4] Artem'ev, Y. M.; Ryabchuk, V. K. *Introduction to Heterogeneous Photocatalysis*; St.Peterburg State Univ.: St.Petersburg, 1999; 304 p. [in Russian].

[5] Ambrus, Z.; Mogyorósi, K.; Szalai, Á.; Alapi, T.; Demeter, K.; Dombi, A.; & Sipos, P. *Appl. Catal. A*. 2008, *vol 340, no 2*, 153-161.

[6] Kiselev, V. F.; Krylov, O. V. *Adsorption and Catalysis on Transition Metals and Their Oxides*; Springer-Verlag: Berlin & New York, NY, 1989; 445 p.

[7] Gao, X.; Bare, S. R.; Fierro, J. L. G.; Banares, M. A.; & Wachs, I. E. *J. Phys. Chem. B*. 1998, *vol 102, no 29*, 5653-5666.

[8] Atkins, P.; Friedman, R. *Molecular Quantum Mechanics;* Oxford Univ. Press: Oxford, 2005; 573 p.

[9] Aleskovskii, V. B. *Chemistry of Supramolecular Compounds*; St.Peterburg State Univ.: St.Petersburg, 1996; 256 p. [in Russian].

[10] Marfunin, A. S. *Physics of Minerals and Inorganic Materials: An Introduction*; Springer-Verlag: Berlin & New York, NY, 1979; 340 p.

[11] Sosnov, E. A.; Malkov, A. A.; & Malygin, A. A. *Integrated Ferroelectrics*. 2009, *vol 103, no 1*, 41-51.

[12] Sosnov, E. A.; Malkov, A. A.; & Malygin, A. A. *Rus J. Phys. Chem. A*. 2009, *vol 83, no 4*, 642-648.

[13] Sosnov, E. A.; Vasil'eva, K. L.; & Malkov, A. A. *Rus J. Phys. Chem. A*. 2010, *vol 84, no 6*, 1028-1032.

[14] Pak, V. N. *Zh Fiz Khim*. 1976, *vol 50, no 5*, 1266-1268.

[15] Pak, V. N.; Tsvetkov, V. K. In *Workshop on the Chemistry of Solids*; Kol'tsov, S. I., Korsakov, V. G., & Smirnov, V. M.; Ed.; Leningrad State Univ.: Leningrad, 1985; pp 160-174. [in Russian].

[16] Pak, V. N.; Ventov, N.G. *Zh Fiz Khim*. 1975. *vol 49, no 10*, 2535-2537.

[17] Kanna, M.; Wongnawa, S. *Mater Chem. & Phys*. 2008, *vol 110, no 1*, 166–175.

[18] Ohtani, B.; Prieto-Mahaney, O. O.; Li, D.; & Abe, R. *J. Photochem. & Photobiology A*. 2010, *vol 216, no 2-3*, 179-182.

[19] Ivanov, A. P. *Optics of Scattering Media*; Nauka i Tekhnika: Minsk, 1969; 592 p. [in Russian].

[20] Klaas, J.; Schulz-Ekloff, G.; & Jaeger N. I. *J. Phys. Chem. B*. 1997, *vol 101, no 8*, 1305-1311.

[21] Gao, X.; Bare, S. R.; Weckhuysen, B. M.; & Wachs, I. E. *J. Phys. Chem. B*. 1998, *vol 102, no 52*, 10842-10852.

[22] Pankove, J. I. *Optical Processes in Semiconductors;* Dover Publications: New York, NY, 1971; 422 p.

[23] Radecka, M. *Materials Science-Poland*. 2006, *vol 24, no 1*, 285-290.

[24] Olthof, B.; Khodakov, A.; Bell, A. T.; & Iglesia E. *J. Phys. Chem. B*. 2000, *vol 104, no 7*, 1516-1528.

[25] Augugliaro, V.; Kisch, H.; Loddo, V.; Lopez-Munoz, M. J.; Marquez-Alvarez, C.; Palmisano, G.; Palmisano, L.; Parrino, F.; & Yurdakal, S. *Appl. Catal. A*. 2008, *vol 349, no 1-2*, 189-197.

[26] Ivanov, A. P.; Loiko, V. A.; & Dik, V. P. *Light Propagation through Close-Packed Disperse Media*; Nauka i Tekhnika: Minsk, 1988; 191 p. [in Russian]

[27] Yavorskii, B. M.; Detlaf, A. A.; & Lebedev A. K. *Handbook on Physics*; Oniks: Moscow, 2006; 1056 p. [in Russian].

[28] Savel'ev, I. V. *Course of General Physics*; Lan': St.Petersburg, 2008; Vol. 3, 320 p. [in Russian].

[29] Janus, M.; Morawski, A. W. *Appl. Catal. B*. 2007, *vol 75, no 1-2*, 118-123.

[30] Chiarello, G. L.; Selli, E.; & Forni, L. *Appl. Catal. B*. 2008, *vol 84, no 1-2*, 332-339.

[31] Noble, S. D.; Boeré, A.; Kondratowicz, T.; Crowe, T. G.; Brown, R. B.; & Naylor, D. A. *Can J. Remote Sensing*. 2008, *vol 34, no 2*, 68-76.

[32] Pak, V. N. *J. Appl. Spectroscopy*. 1975, *vol 22, no 2*, 270-271.

[33] Malygin, A.A. *Rus J. General Chem*. 2002, *vol 72, no 4*, 575-589.

[34] Sosnov, E. A.; Malkov, A. A.; & Malygin, A. A. *Rus. J. Appl. Chem*. 2007, *vol 80, no12*, 2057-2062.

[35] Koshtyal, Yu. M.; Malkov, A. A.; & Malygin, A.A. *Rus. J. General Chem*. 2011, *vol 81, no 1*, 41-48.

[36] Sosnov, E. A. *Rutile: Properties, Synthesis and Applications*; Nova Science Publishers: NY, 2012; 143-158.

In: Optical Spectroscopy
Editor: Nicolae Tomozeiu
ISBN: 978-1-63321-197-1

Chapter 7

POLARISCOPY: ITS HIGH SENSITIVITY TO INTERNAL/RESIDUAL STRAINS OF SEMICONDUCTOR SINGLE CRYSTAL WAFERS

Hideo Takeuchi*

Associate Professor, Ph.D., Department of Applied Physics,
Graduate School of Engineering, Osaka City University,
Sumiyoshi, Osaka, Japan

ABSTRACT

Progress in electronics has been continuously stimulating research and development of various cutting-edge electronic semiconductor devices and components. Characteristics of cutting-edge semiconductor devices essentially rely on the development and improvement of single crystal wafers used for substrates. Internal strains of semiconductor crystals, which reflect imperfections of the crystals, have been conventionally evaluated using photoluminescence spectroscopy and Raman scattering spectroscopy. The referenced spectroscopic evaluations are the characterizing methods with use of observation of such as phonons, carriers, and excitons in solids. Recently, semiconductors, ZnO, GaN, Ga_2O_3, and SiC, have been attracting much attention because their wide band-gap energy enables to realize the devices with a high

* E-mail address: takeuchi@a-phys.eng.osaka-cu.ac.jp Tel: +81-6-6605-2174.

breakdown voltage. The inch-size wafers such as SiC, GaN, and ZnO are now commercially available; however, the yield rate is still discussed issues. In the present chapter, focusing on their optical transparency for visible light, we demonstrate the fact that polariscopy, which utilizes the change in light polarization through the photoelastic effects, is highly sensitive to the strain in comparison with the above-mentioned spectroscopic characterization methods: the Raman scattering and photoluminescence measurement. We provide the findings that the polariscopic analysis is quite suitable to the mass production line owing to the ability of realizing high throughput. We demonstrate the comprehensive and systematical study on the SiC and ZnO wafers. We also present the application example of the polariscopic analysis to evaluating a ZnTe crystal, which is an important nonlinear optical crystal in the field of terahertz spectroscopy. We show effectiveness in the characterization of the strain, comparing the polariscopic analysis and the Raman scattering measurement on the ZnTe crystals. We reach a conclusion that the polariscopic analysis has the internal strain sensitivity higher than the Raman scattering measurement. We also quantitatively present the strain-detection limit of the polariscopic analysis.

I. INTRODUCTION

Wide band-gap semiconductors, which are well regarded as important materials for optoelectronics, have also been receiving much attention in the field of power electronic devices and components because of their advantages in realizing a high breakdown voltage. This is because, for example, the wide band-gap energy suppresses avalanche effects. The avalanche process occurs in the case where the carriers are accelerated by the electric field and obtain the kinetic energy larger than the band-gap energy. The accelerated electrons extremely excite the electron-and-hole pairs via collisions. This leads to the break of the lattice. Consequently, the usage of the wide band-gap semiconductor is a solution to suppress the avalanche effects. In addition, it is self-evident that electric power P is given by the product of current I and voltage V: $P = I \times V$. Accordingly, the operation of the devices, which have a high breakdown voltage, reduces current consumption. In operating a device at the high power condition, conventional devices, which suffer from relatively low breakdown voltage, are connected in parallel. The parallel connection has several issues. The consumption loss caused by each semiconductor device is relatively high in converting the power because of joule heating due to the large current. In addition, each element unbalance in the device gives rise to difficulty in designing electric circuits and units. In order to avoid overheating at the operation, the cooling systems have a large size. Furthermore, the large current operation requires multiplicity of the cables, which leads to large gross weight of the power unit. Effects of cross currents are not negligible. A large number of multiple finger gate structures are required in conventional metal-oxide-semiconductor field effect transistors (MOSFETs) and so on.

Recently, development of electric vehicles becomes vigorous owing to demands for environment protection. In addition, high performance radar systems with a small size, which are sometimes equipped in flying objects and vessels, are desired. In order to realize the size down, it is essential to employ the power devices with a high breakdown voltage. This background accelerates research and development of wide band-gap semiconductors.

Nowadays (in 2014), SiC-based electric devices, for example, metal semiconductor field effect transistors (MESFETs) and Schottky barrier diodes, proceed into the commercial status [1, 2] and their improvement is continuously investigated. $Al_xGa_{1-x}N$/GaN heterostructure-based high electronic mobility transistors (HEMTs) [3], which have the ability of high frequency-voltage operation, are also developed into the mass production status. In fact, the $Al_xGa_{1-x}N$/GaN HEMTs are sometimes used for the radar systems. Several researches cope with Ga_2O_3-based [4], diamond-based [5], and ZnO-based electronic devices [6]. It should be mentioned that, in the SiC-based electronic devices, the effects of the heating are reduced owing to the high heat conductivity.

In the mass production of the semiconductor wafers used for the substrates, it is essential to increase a yield rate; however, the issues on the yield rate are not completely solved. The issues result from the fact that growth technologies for the wide band-gap semiconductor wafers quite differ from those of traditional GaAs, InP, and Si wafers. The GaAs, InP, and Si wafers are grown using liquid-to-solid phenomena. The technique based on the liquid-to-solid phenomena is well established. In contrast, the growth technologies for the wide band-gap semiconductor wafers are state-of-the-art, comparing with the wafer-growth technologies for GaAs, InP, Si, and so on. The SiC wafers are made with use of sublimation methods [7, 8]: the crystallization from the gas phase to solid phase. The GaN wafers are fabricated by hydride vapor-phase epitaxy [9, 10]. For the growth of the ZnO wafers, a hydrothermal method is employed [11]. The referenced growth technologies are the emerging wafer-growth methods; namely, the growth technologies for the wide band-gap semiconductor wafers do not reach a well-established status. It is, therefore, necessary to control quality of the mass production of the wafers. We note the fact that the growth of the semi-insulating wafers, which are indispensable for the HEMTs, requires the relatively high technology. The semi-insulating SiC wafers now become commercially available; however, the growth of the SiC semi-insulating wafers still suffers from suppressing various defects and micro-pipes. It should be taken it account that SiC crystals have polymorphism [12], which causes the difficulty in the growth of the wafers with suitable uniformity. The above-described background indicates that it is significant to explore the reasonable and suitable characterization methods for the mass production. It is, of course, fundamental that the characterization is nondestructive.

In the present chapter, we focus our attention on exploring the convenient and high throughput characterization/inspection technology for internal and/or residual strains of the wide band-gap semiconductor wafers because the relatively large strains sometimes lead to the fragility and resulting cracks. Accordingly, it is essential to suppress the crack in obtaining the high yield rate. In addition, the internal strain usually is concerned with the crystal misorientation of the wafers, on which epitaxial layers are grown. There have been various characterizing technologies for evaluating the internal strains such as photoluminescence spectroscopy and Raman scattering spectroscopy [13, 14]. The referenced spectroscopic characterization has been widely utilized in the research and development of the crystal growth. In addition, the x-ray diffraction measurement and x-ray topography [15] are also regarded as the useful characterizations. On the other hand, we wonder whether the above-mentioned evaluation measurements satisfy the demand from the wafer mass production because these measurements require the relatively long time for the evaluation, leading to the low throughput. The wide band-gap semiconductors are transparent to the visible light, so that the analysis of the transmittance light has the ability of the precise strain evaluation. Polariscopy [16-18], which utilizes photoelastic effects resulting from the stains, is a measuring method on the basis of the fundamental characteristics of light: polarization. In the inhomogeneous materials where the stain effects are dominant, the light which transmits from the sample under the test, exhibits the optical birefringence owing to formation of fast and slow axes. This causes the change in the direction of the polarization. Historically, the polariscopic measurement gives successful and progressive results in the research field of mineralogy. Accordingly, the polariscopic measurement is applicable to the strain evaluation. As we discuss the following sections, the polariscopic analysis has a high sensitivity, compared with the Raman scattering and photoluminescence measurements.

We organize the present chapter with the following sections. We explain the principle of the photoelastic phenomena and the experiment setup of the polariscopic measurements. Next, as an example, we demonstrate the results of the polariscopic analysis for a (0001)-oriented 6H-SiC wafers [19, 20]. We discuss the effectiveness of the polariscopic measurement, comparing with that of the transmittance geometry x-ray topographic measurement. Next, we show the result obtained from the application of the polariscopic measurement to a (0001)-oriented ZnO wafer [21-23]. The elements of Ga and Zn have a high x-ray absorption coefficient [15], so that the conventional transmittance-type x-ray topographic measurement is inappropriate to the wafers of the Ga-*X* or Zn-*X* crystals, where *X* denotes other elements: oxygen, nitrogen and so on. We, at first, employ the reflection-geometry x-ray topographic measurement, which is a specialized and complicated method. We compare the result of the polariscopic measurement with the result of x-ray topographic measurement. In addition, we present the Raman scattering analysis of the ZnO wafer. We also compare the strain sensitivity of the polariscopic analysis with that of the photoluminescence characterization. Finally, we present the results of the polariscopic measurement of a ZnTe single crystal chip [24]. ZnTe crystals are the typical optical nonlinear crystals are used for the detection and emission in terahertz electromagnetic waves enabling the novel optoelectronics [25], so that it is meaningful to clarify the ability of the polariscopic analysis for the ZnTe crystals. We discuss, in detail, the strain sensitivity of the polariscopic analysis, comparing with photoluminescence and Raman scattering measurements.

II. Principle of the Photelastic Effects and the Experiment Setups for the Polariscopic Measurement

The fundamental concepts of the photoelastic effects and the experiment setups for the polariscopic measurements are shown in Figures 1(a) and 1(b). As shown in Figures 1(a) and 1(b), the polarizing plates which are used for the polarizer and analyzer, are in the perpendicular alignment, the cross-Nicol configuration (sometimes referred as cross-Nicoled configuration) [16-18]. The sample under the test is inserted between the polarizer and analyzer. Figure 1(a) schematically shows the case where the sample free from the strains is under the test. In the condition shown in Figure 1(a), the sample is optically isotropic because the symmetry of the crystal does not change. The direction of the polarization of the light is kept, and the light cannot transmit from the analyzer. Figure 1(b) shows the sample that contains the strains. The strains lower the crystal symmetry. The lowering of the symmetry is explained in terms of the fact that the lattice constant is deviated from that of the ideal crystal.

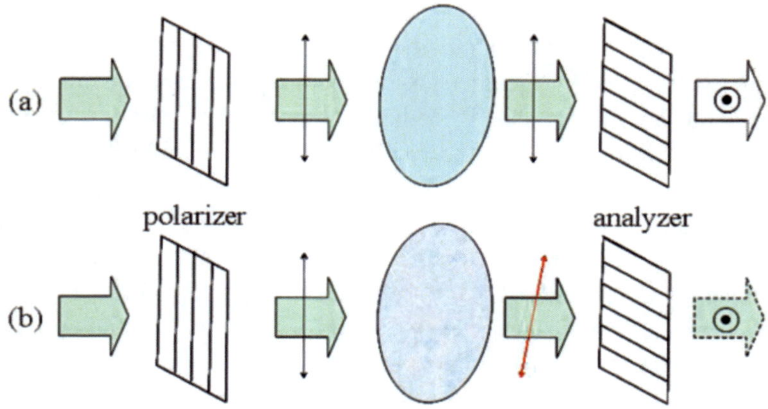

Figure 1. Explanation of the photoelastic phenomenon and concept of the polariscopic measurement.

The above-mentioned phenomenon produces the optical fast and slow axes: the photoelastic effects [16-18]. The production influences the polarization of light and changes the direction, so that the some part components of the light transmitted from the sample is allowed to pass through the analyzer. This is observed as the distributed brightness in the transmittance image. In usual, for the isotropic crystal with a symmetry T_d (F43m), linear polarization optics is used. In the crystal with a symmetry such as C_{6v}^4 ($P6_3mc$), the polariscopic measurement is applicable with use of the left and right circular polarization optics, of the evidence which is shown in Sections III and IV.

Figure 2 shows the example of the setup for the polariscopic measurement in the cross-Nicol configuration. The experiment apparatus is quite simpler than those used for other crystal characterization methods. The light source is a white light-emitting-diode panel. The image, reflecting the intensity of the light transmitted through the analyzer, is recorded with use of a conventional digital still camera. The sample is inserted between the polarizer and analyzer. The polarizing plates used for the polarizer and analyzer are conveniently available Polaroid films. We present the usefulness and sensitivity of the present apparatus in the following chapters.

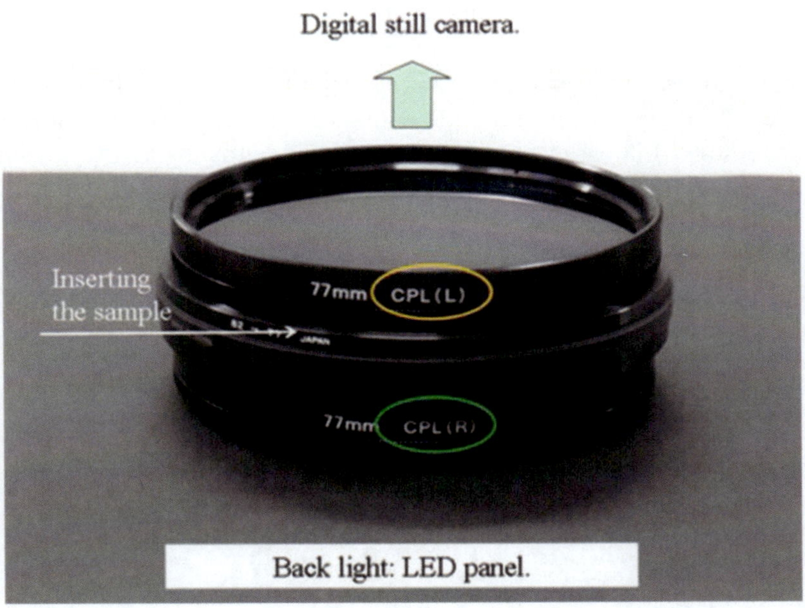

Figure 2. Experimental setup for the polariscopic measurement.

III. Application of Polariscopic Measurement to a Semi-Insulating (0001)-Oriented 6H-SiC Wafer

Figure 3(b) shows the transmittance-type monochromatic x-ray topograph (Lang method) of the semi-insulating 6H-SiC wafer sample. The present sample was a semi-insulating (0001)-oriented 6H-SiC wafer with a diameter of 2 inches and with a thickness of 390 μm. The sheet resistance was higher than 10^5 Ω per square at least. The optical photograph of the sample is shown in

Figure 3(a), in which the optical transparency is uniform. The monochromatic x-ray topographic analysis was performed in the [10-10] reflection geometry. The x-ray used was the Mo Kα1 line with a wavelength of 0.07093 nm. The scan speed and slit width were 1 mm/s and 1 mm, respectively. In the present x-ray topographic measurement, the recording time was 18 minutes. In Figure 3(b), the wafer edge is denoted by the dotted circle.

The undulation of the x-ray topograph, which is recorded as a change in brightness, reflects the disturbance of the magnitude of the crystal-axis orientation. The region, which is dark, corresponds to the low orientation. This means the occurrence of the misalignment. In general, the crystal quality of semiconductor wafers around the center is better than the crystal quality near the edge. The x-ray topograph image extinguishes near the wafer edge. The dark patterns exist in various regions. Consequently, it is impossible to obtain the information on the condition near the wafer edge.

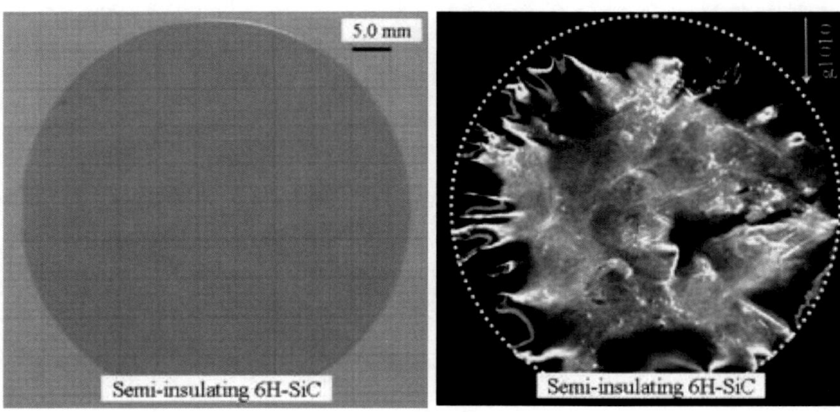

Figure 3. (a) Optical photograph (a) and (b) Transmittance-type x-ray topograph of the present semi-insulating (0001)-oriented 6H-SiC wafer. In Figure 4(b), the undulation is pronounced. Copyright (2011) by the American Institute of Physics.

Next, we performed the circular polariscopic measurement. Here, we describe the advantage of utilizing the circular polarizing plates. The linearly polarized light wave has the characteristics that the direction of the amplitude does not change as a function of time.

In uniaxial crystals like 6H-SiC, the linearly polarized light wave, which passes through a certain light path in the crystals, conserves the original amplitude direction, regardless of the photoelastic effect. In the cross-Nicol configuration, the above-mentioned light wave does not transmit from the analyzer and hence the polariscopic map shows a black cross pattern, the so-called isogyre [16-18]. The appearance of the isogyre brings the difficulty in evaluating the strain distribution in the whole wafer. In the circularly polarized light wave, the amplitude direction rotates with respect to time. The light wave, therefore, transmits the circular analyzer, which enables to obtain the time-integration image.

As a result, the circular crystal like SiC with symmetry of C_{6v}^{4} [26], it is reasonably expected that the circular polariscopic wafer map is not affected by the isogyre. Accordingly, even for the uniaxial polariscopic measurement is applicable to evaluate the strain distribution in the wafer-size area.

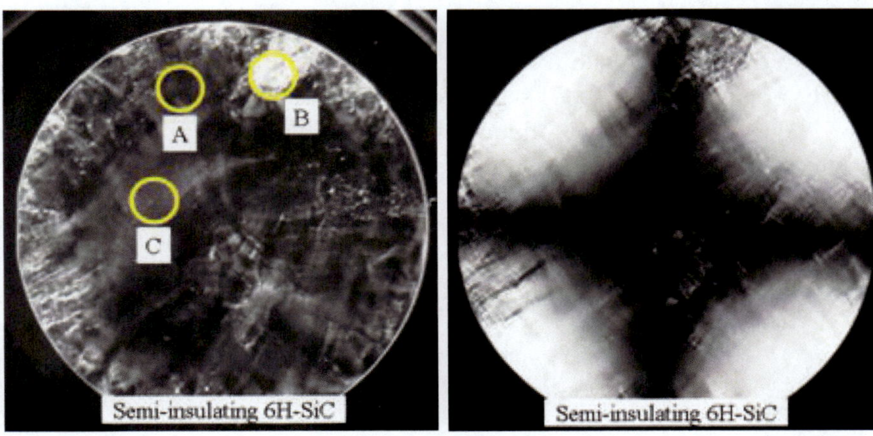

Figure 4. (a) Circular polariscopic map of the sample. (b) Linear polariscopic map of the sample. Each wafer map was taken with use of a digital still camera within 1 sec. The black cross pattern appearing in Figure 4(b) is the so-called isogyre. Copyright (2011) by the American Institute of Physics.

The circular polariscopic map of the sample is shown in Figure 4(a). The linear polariscopic map of the sample is also shown in Figure 4(b), in order to reveal the fact that the appearance of isogyre is annoying. As shown in Figure 4(b), the isogyre is strongly pronounced as the black cross pattern. This indicates that the polariscopic measurement with use of the linear polarizing plates is inappropriate for mapping the strain distribution.

Comparing with Figure 4(b), the circular polariscopic map clearly reflects the condition of the whole wafer as the brightness corresponding to the magnitude of the strain. The present circular polariscopic map was obtained within 1 sec, so that the present method has a high potential to conveniently evaluate the strain distribution in comparison with the x-ray transmittance topographic measurement.

Thus, the circular polariscopic analysis has the ability to complement the monochromatic x-ray topographic analysis applied to the SiC wafer mass production. In order to clarify the origin of the strain, we measured the θ-2θ x-ray diffraction pattern, using the Cu Kα1 line with a wavelength of 0.154 nm. Figure 5 shows the x-ray diffraction patterns of regions A and B indicated in Figure 4(a).

Each x-ray reflection line is assigned using the 6H-SiC lattice constant along the c-axis direction of 1.5112 nm [26]. The line labeled by 000ℓ originates from the (000ℓ) reflection line of 6H-SiC. It is noted that the crystal structure of 6H-SiC is hexagonal. Consequently, according to the extinction rule, the reflection line with an odd value of ℓ is forbidden in ideal crystals perfectly oriented to the [0001] direction [15]. The diffraction pattern from region A does not exhibit the forbidden line. In regions B and C, the forbidden reflection line is observed. Furthermore, the (0006) reflection line is remarkably asymmetric. These phenomena originate from the misaligned crystal blocks and polycrystalline grains, leading to the deviation from the ideal crystal.

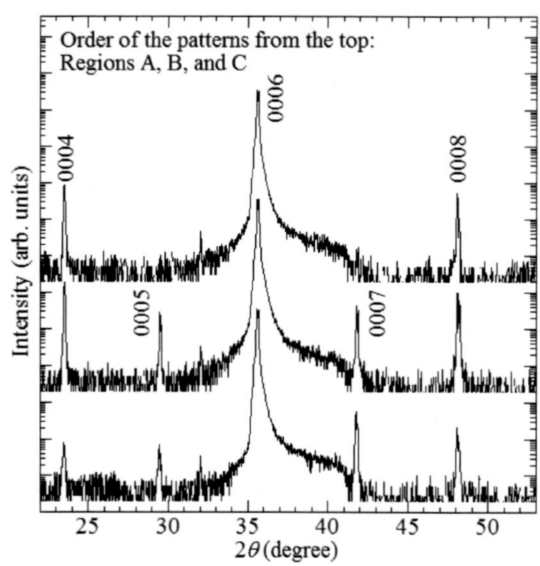

Figure 5. θ-2θ x-ray diffraction pattern at points A, B, and C. The label 000ℓ corresponds to the (000ℓ) reflection line. Copyright (2011) by the American Institute of Physics.

Taking account of the appearance of the forbidden line, the major factor of misaligned crystal blocks and polycrystalline grains is attributed to the phenomenon that the crystal axes of the present sample strongly deviate from the [0001] direction in regions B and C.

This means the offsets of the angle of the orientation along the [0001] direction. We note that the intensity of the forbidden reflection line is weaker by a factor of 10^3 than that of the allowed reflection line. Thus, the polariscopic measurement has the high sensitivity to the strain induced by the misaligned crystal blocks and polycrystalline grains.

IV. Application of Polariscopic Measurement to a (0001)-Oriented ZnO Wafer: Conventionality against the x-ray Topographic Measurement for the Semiconductors with High X-Ray Absorption Elements

Following the comprehensive results obtained in the (0001)-oriented semi-insulating 6H-SiC wafer, we proceed in a next case study in order to confirm the high strain sensitivity of the polariscopic measurement. We, here, demonstrate the experimental results for the (0001)-oriented ZnO wafer.

The present sample was a (0001)-oriented ZnO wafer with a diameter of 2 inches grown by a hydrothermal method. The optical photograph of the sample is shown in Figure 6. The sample has a uniform optical transparency. The monochromatic x-ray topograph was carried out in the reflection configuration (the so-called Berg-Barrett method) [15]. We, for a reference, performed the x-ray topographic measurements in transmittance configuration. We chose an appropriate reflection plane for each x-ray topographic measurement. In the reflection-type x-ray topographic measurement, the applied x-ray line, diffraction plane, and incidence slit width were the Cu Kα line, (202-5) plane, and 2 mm, respectively. The transmission-type x-ray topograph was measured with use of the Lang method in the [101-0] reflection geometry. The x-ray used was the Mo Kα line. The slit width and scan speed were 1 mm and 1 mm/s, respectively.

The transmittance and reflection type monochromatic x-ray topographic images of the sample are shown in Figures 7(a) and 7(b). It is apparent that the transmittance monochromatic x-ray topograph is strongly noisy. The phenomenon originates from the fact that the element of Zn has a high x-ray absorption coefficient [15]. This suppresses the x-ray transmittance and, as a result, results in the noisy image. In contrast, the reflection-type x-ray topograph has the clear image without noise as shown in Figure 7(b), in which the undulation appears. Since the inhomogeneous brightness of the x-ray topograph relating to the disturbance magnitude of the crystal-axis orientation, the emergence of the undulation suggests the appearance of the region misaligned from the [0001] direction. We, as mentioned in Introduction, point out the fact that the x-ray topographic measurement is a specialized and complicated method. Consequently, the reflection-type x-ray topographic measurement is not suitable for the inspection of the ZnO wafers in the mass production.

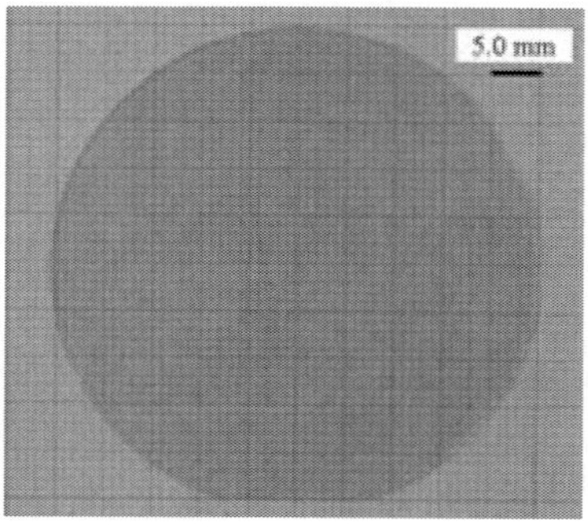

Figure 6. Optical photograph of the present (0001)-oriented ZnO wafer. Copyright (2011) by the American Institute of Physics.

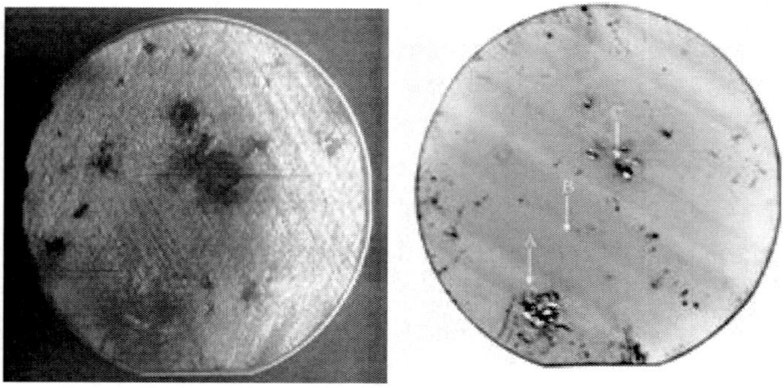

Figure 7. (a) Transmittance-type x-ray topograph taken at the [101-0] reflection geometry. The noisy strongly appears in the whole region. (b) Transmittance-type x-ray topograph taken at the [202-5] reflection geometry. Copyright (2011) by the American Institute of Physics [Figure 7(b)].

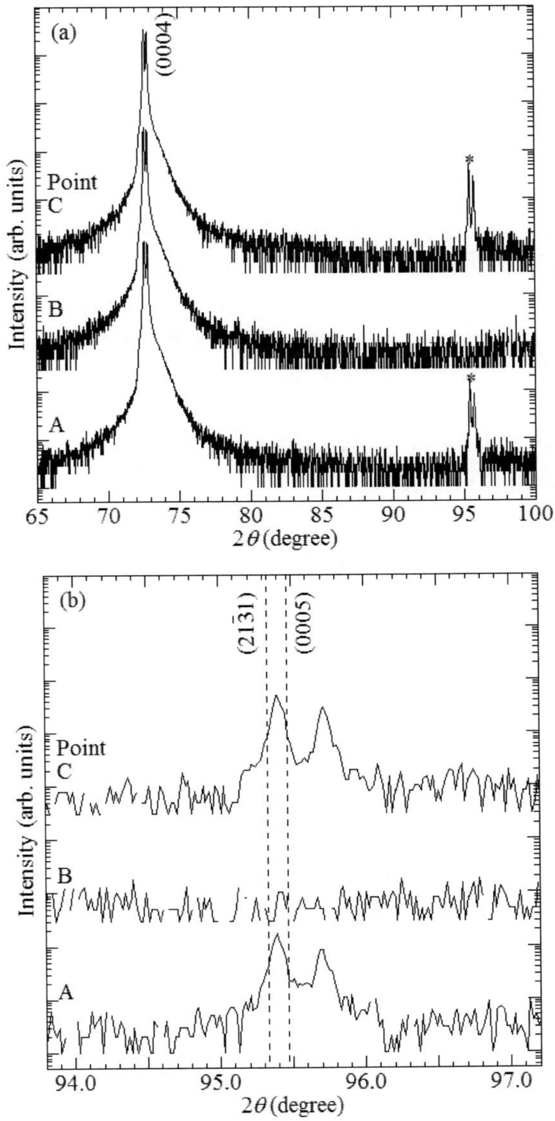

Figure 8. (a) θ-2θ x-ray diffraction patterns of points A, B, and C of the reflection-type x-ray topograph [Figure 7(b)] in the 2θ range from 65 to 100 degrees. For clarity, each diffraction pattern is shifted vertically. (b) θ-2θ x-ray diffraction patterns of points A, B, and C in the 2θ range from 94 to 97 degrees. Copyright (2011) by American Institute of Physics [Fig. 8(a)].

In order to elucidate the origin of the undulation pattern in the x-ray topograph, we performed the θ-2θ x-ray diffraction measurement. The θ-2θ x-ray diffraction patterns were measured using the Cu Kα line. The 2θ range was from 65 to 100 degrees. The accuracy of the 2θ angle was 0.03 degrees. Figure 8(a) shows the θ-2θ x-ray diffraction patterns of regions A, B, and C in the x-ray topograph. Each x-ray reflection line is assigned using the structure results from the reflection of the Cu Kα1 and Cu Kα2 lines in order of increasing 2θ. The ZnO crystals have crystal symmetry of C_4^{6V} ($P6_3mc$) [27]; therefore, the reflection line with an odd value of ℓ is forbidden according to the extinction rule. The main reflection lines observed at points A, B, and C are assigned to the (0004)-reflection line.

The position of each (000ℓ)-reflection line is the same of 72.56 degrees. In region B, the x-ray diffraction pattern exhibits only the (0004)-reflection line. This means that the crystal quality is ideal in region B. In contrast, the diffraction patterns in points A and C, which contains the grain-like patterns in the x-ray topograph, exhibit the reflection line labeled by the asterisks in addition to the (0004)-reflection line. In order to highlight the reflection line labeled by the asterisks, the x-ray diffraction patterns in the range from 94 to 97 degrees are shown in Figure 8(b). The dashed lines correspond to the calculated positions of the (21-31) and (0005) reflections for Cu Ka1 line. The position of the observed peak is located between the (21-31) and (0005)-reflection-line position. Consequently, the observed reflection line in the range from 94 degrees to 97 degrees is assigned the (21-31) and (0005)-reflection lines. The (0005) reflection is forbidden according to the selection rule. In contrast, (21-31) reflection is allowed [15, 28]. The observed reflection line in the range from 94 to 97 degrees mainly contains the (21-31)-reflection component. Consequently, the origin of the grain-like pattern in the x-ray topograph is attributed to the presence of the misaligned crystal blocks and polycrystalline grains.

Next, we performed the circular polariscopic measurement in order to investigate whether the internal stain produces the stress. In the bright patterns of a polariscopic wafer map, a refractive-index change is enhanced by the stresses through the photoelastic effect. The circular polariscopic map of the sample is shown in Figure 9. Regions A and C, which exhibit grain-like patterns in the x-ray topograph, are bright.

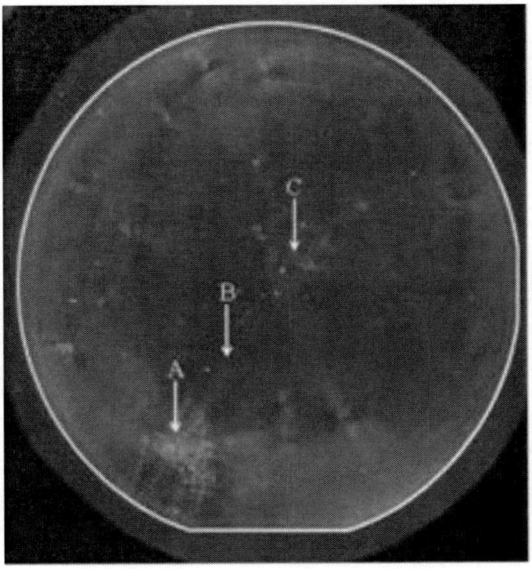

Figure 9. Circular polariscopic map of the sample. Copyright (2011) by American Institute of Physics.

In contrast, region B is dark. The phenomena, which are observed both in the x-ray topograph and in the circular polariscopic map, indicate that the crystal quality is lowered in regions A and C. Taking account of the fact that the intensity of the x-ray reflection line, which appears from 94 to 97 degrees, is weaker by a factor of 10^4 than that of the reflection line at 72.56 degrees, the polariscopic measurement has a high sensitivity. The present findings indicate that the high sensitivity of the polariscopic measurement is universal.

Following the circular polariscopic measurement, we performed the photoluminescence measurement in order to survey the effects of the strains on the luminescence characteristics. The photoluminescence measurement was carried out at room temperature with use of a micro-photoluminescence measurement system. The photon energy and power of the excitation beam were 3.81 eV and 19 mW, respectively. The diameter of the excitation beam was 20 μm on the sample surface. Figure 10(a) shows the photoluminescence spectra at points A, B, and C. Two bands appear in each photoluminescence spectrum.

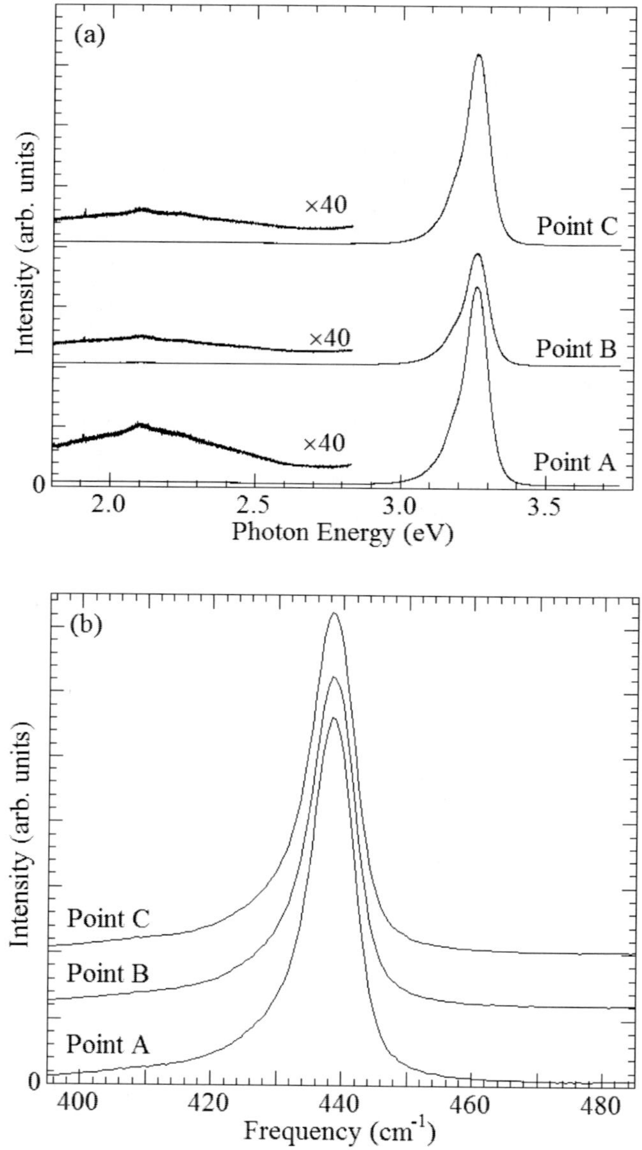

Figure 10. (a) Photoluminescence spectra at room temperature. The broad band peaking at 2.1 eV is attributed to the photoluminescence from the Li or Na impurity deep levels, while The band peaking at the photon energy of 3.26 eV consists of the free exciton photoluminescence and its one-longitudinal-optical-phonon replica. (b) Raman scattering spectra at points A, B, and C at room temperature. The band peaking at 437cm^{-1} is attributed to the E_2^{high} phonon. Copyright (2011) by American Institute of Physics [Fig. 10(a)].

The broad band peaking at 2.1 eV is attributed to the photoluminescence from the Li or Na impurity deep levels [29]. The intensity of the deep level photoluminescence band is the largest in point A.

This suggests an inhomogeneous distribution of the Li or Na impurities. The band peaking at the photon energy of 3.26 eV consists of the free exciton photoluminescence and its one-longitudinal-optical-phonon replica [30-34]. Accordingly, the band at 3.26 eV is assigned to the excitonic photoluminescence band: the band at 3.26 eV reflects the free exciton dynamics in each point.

It is known that the change in the position of the excitonic photoluminescence band connects with various strains. The peaks of the excitonic photoluminescence bands are, however, not shifted; namely, the strains, which are contained in the present wafer, do not have the ability of causing the peak shift of the excitonic photoluminescence bands. According to the difference in the peak intensity, the details are described in Reference 13 in terms of the diffusion of the excitons.

We also measured the Raman scattering spectra of the present sample. In the Raman scattering measurement, the photon energy and power were 2.33 eV and 0.1 mW, respectively. The configuration was (ZXY-Z). Figure 10(b) shows the Raman scattering spectra of the positions A, B, and C in the range from 400 to 480 cm^{-1}. The band peaking at the frequency of 437cm^{-1} is attributed to the E_2^{high} phonon [27]. It is noteworthy that the peak position of the E_2^{high} phonon band is the same of each point; namely, the phonon frequency is not sensitive to the strains that are detected in the polariscopic image map of the present wafer. Next section, we evaluate the strain sensitivity of the polariscopic analysis from the results of the Raman scattering measurement.

V. Polariscopic Measurement of a (110)-Oriented ZnTe Single Crystal Chip: The Estimation of the Strain Sensitivity of the Polariscopic Analysis

In this section, we estimate the strain sensitivity of the polariscopic analysis, using a (110)-oriented ZnTe crystal. The phonon deformation potential of the ZnTe crystal is known, so that the ZnTe crystal is suitable for the estimation of the strain sensitivity.

In the present investigation, we used a (110)-oriented ZnTe crystal chip with the size of 1.0 cm × 1.1 cm. The thickness was 500 μm. The optical photograph of the present sample was uniform as shown in Figure 11(a): The light scattering by the polycrystals was negligible.

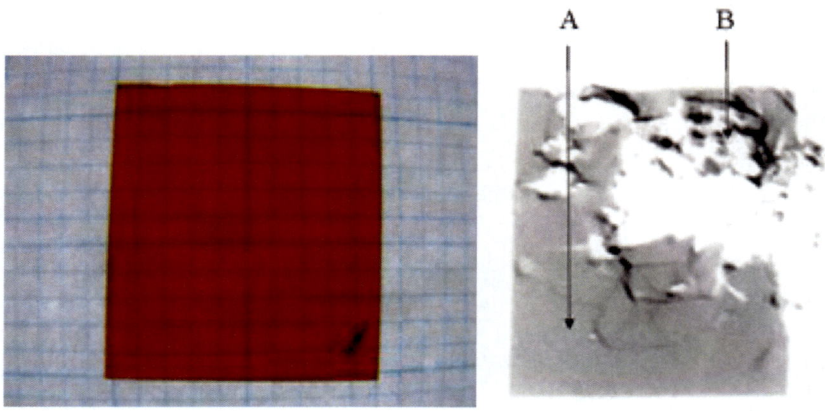

Figure 11. (a) Optical photograph of the present (110)-oriented ZnTe sample. (b) Transmittance-type x-ray topograph recorded at the [531] geometry. Polariscopic image map obtained with use of the linear polarizer and analyzer, using the cross-Nicol configuration. The brightness reflects the intensity of the transmitting light. The pattern appearing the lower right corner is due to the mark.

Figure 11(b) shows the reflection-type x-ray topograph of the present ZnTe sample. The reflection type x-ray topograph was obtained with use of the Cu Kα x-ray line. The diffraction plane was a (531) plane and the slit width was 1.0 mm. It is clear that the undulation is observed. The undulation is considerably large in the upper right corner side. As discussed in the former sections, the emergence of the undulation suggests the presence of the region misaligned from the [110] direction. In order to elucidate the origin of the undulation pattern in the x-ray topograph, we performed the micro θ-2θ x-ray diffraction measurement. The micro θ-2θ x-ray diffraction pattern was measured using the Co Kα line. The 2θ angle ranged from 15 to 80 degrees. Figure 12 shows the micro θ-2θ x-ray diffraction patterns at points A and B in the x-ray topograph. The x-ray diffraction pattern at point A exhibits the reflection lines at the 2θ angles of 44 and 49 degrees. The lattice constant of ZnTe is 0.6102 nm [26], so that the observed lines are attributed to the (220) reflection of the Co Kα1 line and the (220) reflection of the Co Kα line, in order of increasing 2θ angle. The observation of the (220)-reflection line is reasonable because the present sample is oriented to the [110] direction. In the x-ray diffraction pattern at point B, the additional line is detected at 28 degrees. The origin of this line is attributed to the (111) reflection of the Co Kα line. The observation of the (111)-reflection line indicates the formation of the misalignment from the [110] direction. The intensity of the (111)-reflection line is weaker by the factor of 10^2 than that of the (220)-reflection line. This suggests that the misalignment is quite small, which extinguishes the light scattering in the optical photograph of the present ZnTe sample.

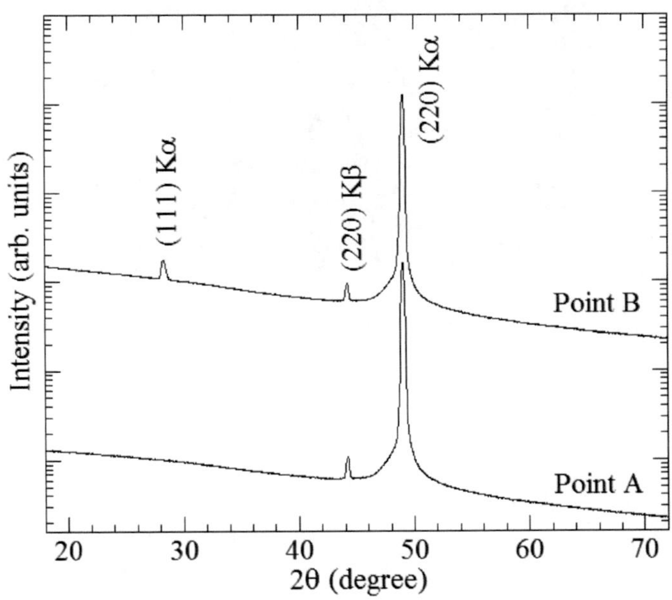

Figure 12. Micro θ-2θ x-ray diffraction patterns at points A and B in the x-ray topograph.

The polariscopic image is shown in Figure 3. In the measurement of the polariscopic image map, for the simplicity, the linear polarizing plates were used because the ZnTe crystal has an isotropic zinc blend structure. As mentioned above, the isogyre does not appear in the isotropic structure. As shown in Figure 13, the inhomogeneous light transmission is clearly observed at the upper right corner region that exhibits in the same position of the undulation pattern in the x-ray topograph as shown in Figure 13. The observed light transmission indicates the presence of the internal strain that induces the refractive index anisotropy.

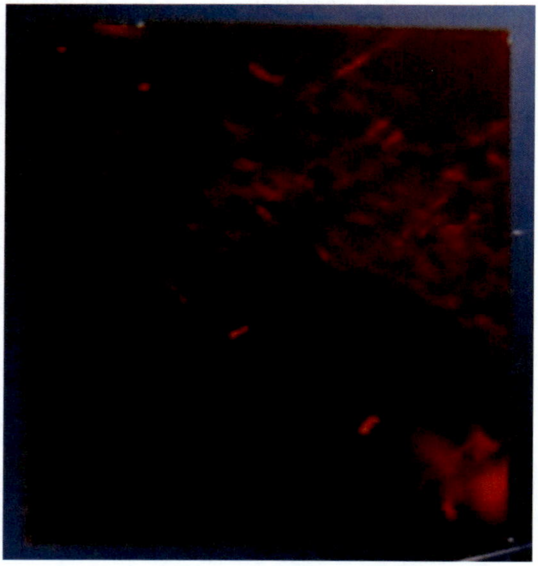

Figure 13. Polariscopic image map obtained with use of the linear polarizer and analyzer at the cross-Nicol configuration. The brightness reflects the intensity of the transmitting light. The pattern appearing the lower right corner is due to the mark.

Following the polariscopic analysis, we carried out the Raman scattering measurement at room temperature. The Raman scattering measurement was carried out in the back scattering configuration. The photon energy and power of the excitation laser beam were 1.96 eV and 0.7 mW, respectively. The spectral resolution power was 0.4 cm^{-1}. Taking account of only the

Raman selection rule, which means the off-resonant condition, only the transverse optical phonon (TO) is allowed. The photon energy, which was chosen in the present work, however, causes the appearance of longitudinal optical (LO) phonon band through the Fröhlich interaction [5, 36, 37] related to the resonance effects; namely, the LO and TO phonon bands can be simultaneously observed in the present Raman scattering measurement. Since the deformation potential of the LO phonon is known, it is reasonable to observe the LO phonon for estimating the magnitude of the strains. The Raman scattering spectra at points A and B are shown in Figure 14(a). Each Raman scattering spectrum exhibits the bands at the frequencies of 177 and 206 cm^{-1}. The Raman bands are assigned to the TO and LO phonons in order of increasing frequency [38]. The LO phonon band, which is forbidden in the present configuration for (110)-oriented zinc blende structure crystals, actually

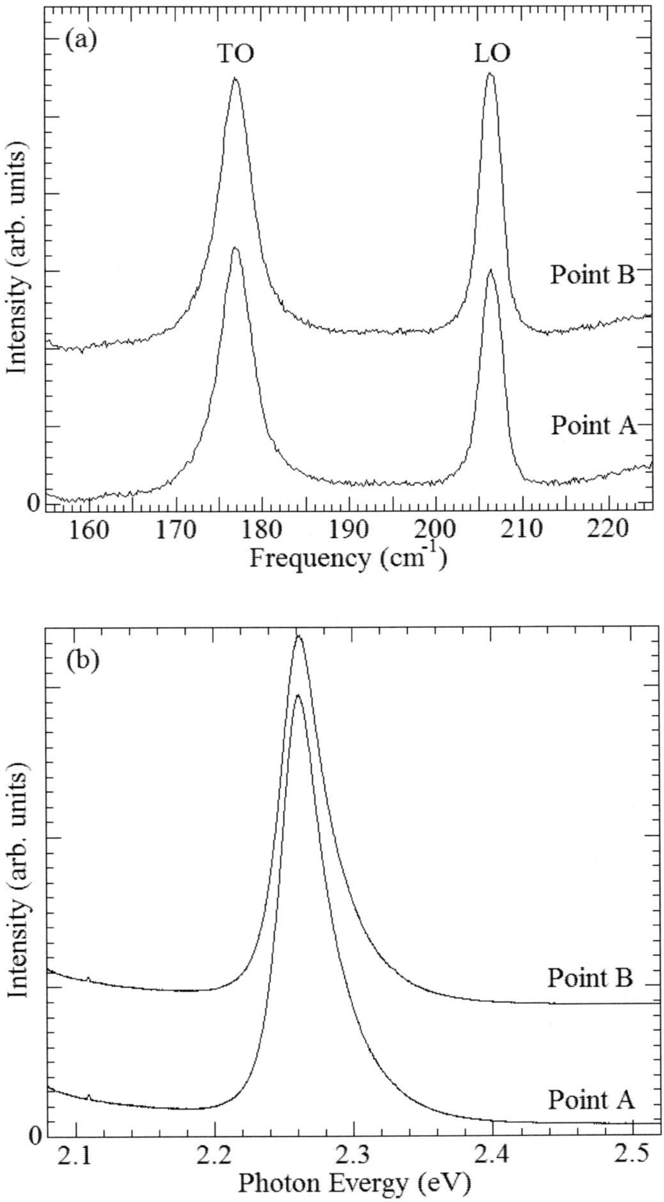

Figure 14. (a) Raman spectra at points A and B in the x-ray topograph shown in Fig. 1. The Raman bands labeled by TO and LO originates from the transverse optical (TO) and longitudinal optical (LO) phonons, respectively. (b) Photoluminescence spectra at positions A and B. The band peaking at the photon energy of 2.26 eV originates from the interband transition.

appears owing to the Fröhlich interaction. As shown in Figure 14(a), the peak positions of the LO and TO phonon bands at each point are the same within the spectral resolution power, indicating the quite low sensitivity of the Raman scattering measurement to the internal strain of the ZnTe crystal. Thus, it is concluded that the polariscopic analysis has the internal strain sensitivity higher than Raman scattering measurement.

From now on, we discuss the highest limit of the strain sensitivity. Equation (1) expresses the frequency shift $\Delta\omega$ of the LO phonon, which is caused by the strain ε: [36, 38]

$$\frac{\Delta\omega}{\omega} = \left(K_{12} - K_{11} \frac{C_{12}}{C_{11}} \right) \varepsilon \tag{1}$$

Here, the physical quantity ω denotes the LO phonon frequency. The parameters K_{11} and K_{12} are related to the phonon deformation potentials, while C_{11} and C_{12} are the elastic constants. In the ZnTe crystal, the K_{11} (K_{12}) value is -2.3 (-1.7), and the ratio C_{11}/C_{12} is 0.574 [38]. Taking account of the spectrum resolution of 0.4 cm^{-1}, the strain-sensitivity limit is 0.5% at most in the Raman scattering measurement. The present spectral resolution is general in performing Raman scattering measurements. As mentioned above, the internal strain, the effect of which does not shift the phonon frequency, is detectable in the polariscopic analysis. Accordingly, we conclude that the polariscopic analysis detects the internal strain less than 0.5%.

Finally, we present the photoluminescence characteristics in order to confirm that the photoluminescence evaluation is also insensitive to the strain. Figure 14(b) shows the photoluminescence spectra at points A and B. The photoluminescence measurement was performed at room temperature with use of a micro-photoluminescence measurement system. The photon energy and power of the excitation beam were 2.81 eV and 0.1 mW, respectively. The diameter of the excitation beam was 20 μm on the sample surface. In each photoluminescence spectra, the band appears at the photon energy of 2.26 eV [26]. Since the band-gap energy of ZnTe is 2.25 eV, the observed band is attributed to the interband transition. The peak energy of the observed photoluminescence band is the same between points A and B. Thus, we conclude that the polariscopic analysis is also more sensitive than the photoluminescence measurement.

CONCLUSION

We have systematically investigated the feasibility of the polariscopic analysis of the strains contained in the wide-band gap semiconductor wafers, the 6H-SiC wafer and the ZnO wafer, together with the ZnTe chip-size sample. We have also used the reflection-type and transmittance type x-ray topographic measurements. We have observed the undulation in the x-ray topograph, the position of which contains the misalignment the ideal orientation. The presence of the misalignment has been confirmed from the x-ray diffraction pattern. The polariscopic analysis has clarified that, in the undulated region of the x-ray topograph, the light incident to the sample inhomogeneously transmits from the analyzer. The light transmission indicates that the refractive index anisotropy is produced by the internal strain. We have presented the optical properties of the tested sample. We have investigated their relation with the strain detected using polariscopy. The Raman scattering measurement has been revealed to be insensitive to the internal strain, compared with the strain detected using polariscopy. The photoluminescence characteristics have also been insensitive. We, therefore, conclude that the polariscopic analysis has the internal strain sensitivity higher than the Raman scattering and photoluminescence measurements. From the strain-detection limit of the Raman scattering measurement, we have deduced that the polariscopic analysis detects the strain less than 0.5%.

To finalize the present paper, we briefly describe the advantage of the polariscopic analysis against the x-ray topograph. Since the elements, Ga and Zn, have high x-ray absorption coefficients, the widely-used transmission-type x-ray topographic measurement lacks the capability of evaluating the compound semiconductors containing Ga or Zn cation. This connects with the reason why we utilized reflection-type x-ray topographic measurement; however, the reflection-type measurements are generally accompanied with un-conventionality and difficulty. The x-ray topographic analysis is categorized into the qualitative evaluation methods. The polariscopic analysis is, therefore, the best selection, in the case where the requirement for the evaluation is limited to be qualitative. We note the further potential of the polariscopic analysis. At present, it is hard to obtain detailed data on the strain-induced refractive index anisotropy. The polariscopic analysis is, therefore, qualitative at the current status. It is, on the other hand, also worthy pointing it out that the detailed investigation on the strain-induced refractive index anisotropy realizes the quantitative strain estimation.

REFERENCES

[1] S. Dimitrijev, H. B. Harrison, P. Tanner, K. Y. Cheong, and J. Han, Oxidation, MOS Capacitors, and MOSFETs in: *SiC Power Material: Device and Applications* edited by Z. C. Feng (Springer-Verlag, Berlin, Heidelberg, 2004) Chapter 1 and the references therein.

[2] Z. Luo, T. Chen, D. C. Sheriedan, and J. D. Cressler, 4H-SiC Power-Switching Devices for Extreme-Environment Applications in: *SiC Power Material: Device and Applications* edited by Z. C. Feng (Springer-Verlag, Berlin, Heidelberg, 2004) Chapter 1 and the references therein.

[3] J.-Y Duboz and M. A. Khan, Transistors and Detectors Based on GaN-related materials in: *Group III Nitride Semiconductor Compounds: Physics and Applications* edited by B. Gil (Oxford, 1998) Chapter 9.

[4] M. Higashiwaki, K. Sasaki, A. Kuramata, T. Masui and S. Yamakoshi, *Applied. Physics Letters.* 100, 013504 (2012).

[5] K. Kodama, T. Funaki, H. Umezawa, S. Shikata, *IEICE Electronics Express* 7, 17 (2010).

[6] D. C. Look, *Materials Science and Engineering: B* 80, 383 (2001).

[7] T. Sudarson, Material Science and Engineering of Bulk Silicon Carbides in: *SiC Power Material: Device and Applications* edited by Z. C. Feng (Springer-Verlag, Berlin, Heidelberg, 2004) Chapter 1 and the references therein.

[8] N. Ohtani, T. Fujimoto, M. Katsuno, and H. Yahiro, Sublimation Growth of SiC Single Crystals, in: *SiC Power Material: Device and Applications* edited by Z. C. Feng (Springer-Verlag, Berlin, Heidelberg, 2004) Chapter 1 and the references therein.

[9] Y. Oshima, T. Suzuki, T. Eri1, Y. Kawaguchi, K. Watanabe, M. Shibata, and T. Mishima, *Journal of Applied Physics* 98, 103509 (2005).

[10] Y. Oshima, T. Eri, K. Ikeda, S. Yamamoto, K. Watanabe, M. Shibata, T. Mishima, *Journal of Crystal Growth* 310, 5 (2008).

[11] Waag, Growth in: *Zinc Oxide: From Fundamental Properties towards Novel Applications*, edited by C. F. Klingshirn, M. K. Mayer, A. Waag, A. Hoffmann, and J. Geurts (Springer-Verlag, Berlin, Heidelberg, 2010) Chapter 3 and the references therein.

[12] J. Dong and A.-B. Chen, Fundamental Properties of SiC: Crystal Structure, Bonding Energy, Band Structure, and Lattice Vibrations in: *SiC Power Material: Device and Applications* edited by Zhe Chuang

Feng (Springer-Verlag, Berlin, Heidelberg, 2004) Chapter 2 and the references therein.

[13] M. Cardona, *Light Scattering in Solids II* (Springer-Verlag, Heidelberg, 1982) and the references therein.

[14] Pelant and J. Valenta, *Luminescence Spectroscopy and Semiconductors* (Oxford, 2012).

[15] B. D. Cullity, *Elements of X-Ray Diffraction -2nd Edition* (Addison-Wesley, 1978).

[16] G. Cloud, *Optical Methods of Engineering Analysis* (Cambridge University Press, 1994) Chapter 4.

[17] E. E. Wahlstrom, *Optical Crystallography* (John Wiley & Sons, New York, 1951).

[18] V. P. Kompaneĭtsev, *Crystallography Reports* 51, 640 (2006).

[19] H. Takeuchi, *Rev*iew of *Scientific Instruments*. 82, 033907 (2011).

[20] H. Takeuchi, *Physica Status Solidi C* 9, 266 (2012).

[21] H. Takeuchi, *AIP Advances* 1, 042176 (2011).

[22] H. Takeuchi, *Physica Status Solidi C* 9, 1817 (2012).

[23] H. Takeuchi, *The European Physical Journal B* 86, article no. 50 1-5 (2013).

[24] H. Takeuchi, *Physica Status Solidi*, in press (2014).

[25] Sakai (ed.) *Terahertz Optelectronics* (Springer, Berlin, Heidelberg, 2004) and the refereces therein.

[26] *Semiconductors -Basic Data: 2nd revised edition* edited by O. Madelung (Springer, Berlin, Heidelberg, 1996).

[27] Geurts, Crystal Structure, Chemical Binding, and Lattice Properties in: *Zinc Oxide: From Fundamental Properties towards Novel Applications*, edited by C. F. Klingshirn, M. K. Mayer, A. Waag, A. Hoffmann, and J. Geurts (Springer-Verlag, Berlin, Heidelberg, 2010) Chapter 2 and the references therein.

[28] *International Tables for Crystallography Vol. A* edited by T. Hahn (4th edition, Kluwer Academic Publishers, 1995) pp. 574-575.

[29] Hoffmann, E. Malguth, and B. K. Meyer, Deep Centers in ZnO in: *Zinc Oxide: From Fundamental Properties towards Novel Applications*, edited by C. F. Klingshirn, M. K. Mayer, A. Waag, A. Hoffmann, and J. Geurts (Springer-Verlag, Berlin, Heidelberg, 2010) Chapter 9 and the references therein.

[30] D. W Hamby, D. A. Lucca, M. J. Klopfstein, and G. Cantwell, *Journal of Applied Physics* 93, 3214 (2003).

[31] W. Shan, W. Walukiewicz, J. W. Ager III, K. M. Yu, H. B. Yuan, H. P. Xin, G. Cantwell, and J. J. Song, *Applied Physics Letters* 86, 191911 (2005).

[32] Wang and N. C. Giles, *Journal of Applied Physics* 94, 973 (2003).

[33] S. Yamamoto, H. Sakuma, and T. Mishina, *Japanese Journal of Applied Physics* 49, 121102 (2010).

[34] Yamamoto, Y. Moriwaki, K. Hattori, and H. Yanagi, *Applied Physics Leters* 98, 061907 (2011).

[35] R. Loudon, *Advances in Physics* 13, 423 (1964).

[36] B. Jusserand and M. Cardona, Raman Spectroscopy of Vibrations in Superlattices in: *Light Scattering in Solids V*, M. Cardona and G. Güntherodt (eds.) (Springer, Berlin, 1989) Chapter 3.

[37] H. Fröhlich, *Advance in Physics* 3, 325 (1954).

[38] R. J. Thomas, M. Boley, H. R. Chandrasekhar, M. Chandrasekhar, C. Parks, A. K. Ramdas, J. Han, K. Kobayashi, and R. L. Gunshor, *Physical Reveiw B* 49, 2181 (1994).

INDEX

A

B

C

D

E

F

G

H

I

J

K

L

M

N

O

P

Q

R

S